Strategies for Sustainability

Series Editors
Lawrence Susskind
Ravi Jain

For further volumes:
http://www.springer.com/series/8584

Strategies for Sustainability

Aims and Scope

The series, will focus on "implementation strategies and responses" to environmental problems – at the local, national, and global levels. Our objective is to encourage policy proposals and prescriptive thinking on topics such as: the management of sustainability (i.e. environment-development trade-offs), pollution prevention, clean technologies, multilateral treaty-making, harmonization of environmental standards, the role of scientific analysis in decision-making, the implementation of public-private partnerships for resource management, regulatory enforcement, and approaches to meeting inter-generational obligations regarding the management of common resources. We will favour trans-disciplinary perspectives and analyses grounded in careful, comparative studies of practice, demonstrations, or policy reforms. We will not be interested in further documentation of problems, prescriptive pieces that are not grounded in practice, or environmental studies. Philosophically, we will adopt an open-minded pragmatism – "show us what works and why" – rather than a particular bias toward a theory of the liberal state (i.e. "command-and- control") or a theory of markets.

We invite Authors to submit manuscripts that:

Prescribe how to do better at incorporating concerns about sustainability into public policy and private action.

Document what has and has not worked in practice.

Describe what should be tried next to promote greater sustainability in natural resource management, energy production, housing design and development, industrial reorganization, infrastructure planning, land use, and business strategy.

Develop implementation strategies and examine the effectiveness of specific sustainability strategies. Focus on trans-disciplinary analyses grounded in careful, comparative studies of practice or policy reform.

Provide an approach "…to meeting the needs of the present without compromising the ability of future generations to meet their own needs," and do this in a way that balances the goal of economic development with due consideration for environmental protection, social progress, and individual rights.

The Series Editors welcome any comments and suggestions for future volumes

SERIES EDITORS
Lawrence Susskind
susskind@mit.edu
Professor Ravi Jain
rjain@pacific.edu

Ravi Jain • Luke Lee
Editors

Fiber Reinforced Polymer (FRP) Composites for Infrastructure Applications

Focusing on Innovation, Technology Implementation and Sustainability

Springer

Editors
Ravi Jain
School of Engineering and Computer Science
University of the Pacific
3601 Pacific Avenue,
Stockton, CA 95211
USA
rjain@pacific.edu

Luke Lee
School of Engineering and Computer Science
University of the Pacific
3601 Pacific Avenue
Stockton, CA 95211
USA
llee4@pacific.edu

ISBN 978-94-007-2356-6 e-ISBN 978-94-007-2357-3
DOI 10.1007/978-94-007-2357-3
Springer Dordrecht Heidelberg London New York

Library of Congress Control Number: 2011943093

Printed on acid-free paper

Springer is part of Springer Science+Business Media (www.springer.com)

Preface

This book was made possible by the exceptional support provided by the US Army Engineer Research and Development Center (ERDC), Construction Engineering Research Laboratory (CERL), and University of the Pacific School of Engineering and Computer Science.

The authors, as listed in this book, took the time to prepare excellent manuscripts focusing on various issues related to fiber reinforced polymer (FRP) composites for infrastructure applications. The authors specifically focused on innovation, technology implementation, and sustainability concepts. These manuscripts were rigorously reviewed and refereed by scientists and engineers knowledgeable in the field before inclusion in this book. An introductory chapter was prepared to provide an overview on the main topics and to integrate technical issues covered in the book.

FRP composites are increasingly being considered for major infrastructure components or systems; this is particularly true when there is a need to upgrade, strengthen, or provide seismic retrofit to existing structures. The lightweight, high strength, high stiffness, and corrosion resistant characteristics of FRPs have motivated engineers and scientists to explore their application to complex infrastructure systems. There are, however, crucial issues related to environment and sustainability: theses concepts are covered in this book as well.

The editors are most grateful to the contributors, sponsor organizations, and the many colleagues who were kind enough to assist in making this book possible. We are particularly grateful to Dr. Ashok Kumar, Dr. Larry Stephenson, and Ms. Debbie Lawrence (CERL) for providing general guidance and direction for the book and to Ms. Deanna Thompson and Ms. Mickie Sundberg (University of the Pacific) for assisting with managing organizing and editing the book manuscript.

Background information about the editors and contributors to this book is provided in the following pages.

University of the Pacific, California, USA — Ravi K. Jain and Luke Lee

Contents

Part II Applications

Contributors

Avinash Reddy Akepati Department of Aerospace Engineering and Mechanics, University of Alabama, 209 Hardaway Hall, 7th Ave., Tuscaloosa, AL 35487, USA, avinash@vt.edu

Delicia H. Borja School of Engineering and Computer Science, University of the Pacific, 3601 Pacific Ave. Stockton, CA 95211, USA, d_borja@u.pacific.edu

Lijuan Cheng Department of Civil and Environmental Engineering, University of California Davis, One Shields Ave., Davis, CA 95616, USA, dawcheng@ucdavis.edu

Piyush K. Dutta Dutta Technologies Inc., 4810 Euginea Drive, Palm Beach Gardens, FL 33418, USA, pkdutta@aol.com

Hector Estrada Department of Civil Engineering, University of the Pacific, 3601 Pacific Ave., Stockton, CA 95211, USA, hestrada@pacific.edu

Mohamed A. Faruqi Department of Civil and Architectural Engineering, Texas A & M University-Kingsville, Kingsville, TX 78363, USA, m-faruqi@tamuk.edu

Dean C. Foster Composite & Hybrid Materials Branch, Wright Patterson Air Force Base, Dayton, OH 45433-7750, USA, dean.foster@wpafb.af.mil

Claudia Gonzalez-Liendo Texas Department of Transportation, 1817 Bob Bullock Loop, Laredo, TX 78040, USA, CGONZALE@dot.state.tx.us

Anwarul Haque Department of Aerospace Engineering and Mechanics, University of Alabama, 209 Hardaway Hall, 7th Ave., Tuscaloosa, AL 35487, USA, ahaque@eng.ua.edu

David Holdener Department of Civil, Architectural, and Environmental Engineering, Missouri University of Science and Technology, 325 Butler Carlton Hall, 1401 North Pine Street, Rolla, MO 65409, USA, holdener@umr.edu

Mahesh Hosur Center for Advanced Materials, Tuskegee University, 1200 W. Montgomery Rd., Tuskegee Inst, Tuskegee, AL 36088, USA, hosur@tuskegee.edu

Ravi Jain School of Engineering and Computer Science, University of the Pacific, 3601 Pacific Ave., Stockton, CA 95211, USA, rjain@pacific.edu

Shaik Jeelani Center for Advanced Materials, Tuskegee University, 1200 W. Montgomery Rd., Tuskegee Inst, Tuskegee, AL 36088, USA, jeelanis@tusk.edu

Brent Kawahara Department of Civil Engineering, University of the Pacific, 3601 Pacific Avenue, Stockton, CA 95211, USA, B_kawahara@u.pacific.edu

Ashok Kumar U.S. Army Engineer Research and Development Center, Construction Engineering Research Laboratory (ERDC-CERL), Champaign, IL 61821-9005, USA, Ashok.kumar@usace.army.mil

Richard G. Lampo U.S. Army Corps of Engineers, Engineer Research and Development Center, Construction Engineering Research Laboratory (ERDC-CERL), Champaign, IL 61826-9005, USA, richard.g.lampo@usace.army.mil

Luke Lee School of Engineering and Computer Science, University of the Pacific, 3601 Pacific Ave., Stockton, CA 95211, USA, llee4@pacific.edu

Jennifer K. Lynch Materials Science and Engineering Department, Rutgers University, Piscataway, NJ 08854-8065, USA, jklynch@rci.rutgers.edu

Wesley Merkle Department of Civil, Architectural, and Environmental Engineering, Missouri University of Science and Technology, 325 Butler Carlton Hall, 1401 North Pine Street, Rolla, MO 65409, USA, wmerkle@structurepoint.com

John J. Myers Deparment of Civil, Architectural, and Environmental Engineering, Missouri University of Science and Technology, 325 Butler Carlton Hall, 1401 North Pine Street, Rolla, MO 65409, USA, jmyers@mst.edu

Abilash R. Nair Department of Aerospace Engineering and Mechanics, University of Alabama, 209 Hardaway Hall, 7th Ave., Tuscaloosa, AL 35487, USA, arnair@crimson.ua.edu

Thomas J. Nosker Materials Science and Engineering Department, Rutgers University, Piscataway, NJ 08854-8065, USA, tjnosker@rci.rutgers.edu

Patrick Plews TranSystems, 4555 Lake Forest Drive, Suite 540, Blue Ash, OH 45242, USA, pjplews@transystems.com

Christina Ramirez School of Engineering and Computer Science, University of the Pacific, 3601 Pacific Ave., Baun Hall, Stockton, CA 95211, USA, cramirez@u.pacific.edu

Samit Roy Department of Aerospace Engineering and Mechanics, University of Alabama, 209 Hardaway Hall, 7th Ave., Tuscaloosa, AL 35487, USA, sroy@eng.ua.edu

Joseph O. Sai Department of Civil and Architectural Engineering, Texas A & M University-Kingsville, Kingsville, TX 78363, USA, J-Sai@tamuk.edu

Larry Stephenson US Army ERDC-CERL, 2902 Newmark Dr., Champaign, IL 61826, USA, larry.d.stephenson@usace.army.mil

Kiang Hwee Tan Department of Civil & Environmental Engineering, National University of Singapore, Block E1A, #07-03, 1 Engineering Drive 2, Singapore 117576, Singapore, cvetankh@nus.edu.sg

Jonathan Trovillion US Army Engineer Research and Development Center – Construction Engineering Research Laboratory (ERDC-CERL), PO Box 9005, Champaign, IL 61826, USA, jonathan.c.trovillion@usace.army.mil

Shaik Zainuddin Center for Advanced Materials, Tuskegee University, 1200 W. Montgomery Rd., Tuskegee Inst, Tuskegee, AL 36088, USA, szainuddin@mytu.tuskegee.edu

Manoochehr Zoghi Lyles College of Engineering, California State University, Fresno, 2320 East San Ramon Ave., M/S EE94, Fresno, CA 93740, USA, mzoghi@csufresno.edu

About the Editors

Ravi K. Jain is Dean, School of Engineering and Computer Science, University of the Pacific. Prior to this appointment, he has held research, faculty, and administrative positions at the University of Illinois (Urbana-Champaign), Massachusetts Institute of Technology (MIT), and the University of Cincinnati. He received degrees in civil engineering (BS, MSCE) from California State University. He holds a Ph.D. from Texas Tech University and a Masters in Public Administration from Harvard University. He has directed major research programs for the US department of Defense and has served as chairman of the Environmental Engineering Research Council, ASCE. He is a member of the American Academy of Environmental Engineers, a fellow of ASCE, and a fellow of AAAS. He has been a Littauer fellow at Harvard University and a fellow of Churchill College, Cambridge University. He has published 14 books, over 150 scholarly papers and technical reports, and has received many awards and honors including the US Army's highest research award and was a recipient of the NSPE Founders Gold Medal and named Federal Engineering of the Year for 1989.

Luke Lee, Associate Professor of Civil Engineering, School of Engineering and Computer Science, University of the Pacific, Stockton, California. He received his B.S. in Civil Engineering from the University of California, Los Angeles, his M.S. in Structural Engineering from the University of California, Berkeley and a Ph.D. in Structural Engineering from the University of California, San Diego.

He is well-known for his work on infrastructure renewal using fiber reinforced polymer (FRP) composites, durability characterization of FRP composites, service life estimation, and structural health monitoring of civil infrastructure. His prior appointments include teaching and industry positions at Louisiana Tech University (Ruston, Louisiana) and Structural Integrity Associates, Inc. (San Jose, CA).

Chapter 1
Introduction

Luke Lee, Ravi Jain, Larry Stephenson, and Christina Ramirez

Abstract Over the past several decades the use of fiber reinforced polymer (FRP) composites has gained acceptance in civil infrastructure as a result of the need to rehabilitate or retrofit existing structures, construct infrastructure systems faster, and increase the usable life of the built environment. In addition, increased attention towards a sustainable built environment has challenged engineers to weigh environmental and social impacts in addition to traditional measures of performance and cost of the built environment. As an introduction to the book, this chapter provides background information on FRP composites and identifies the specific challenges associated with the materials and its applicability in a sustainable environment. Next, a unique perspective from the Department of Defense is presented on the challenges and application of FRP composites in sustainable development. Finally, we detail the environmental implications associated with the constituent materials, namely fiber and matrix, and composite manufacturing processes associated with FRP composites in civil infrastructure.

Keywords Fiber reinforced polymer composites • Infrastructure • Sustainability

L. Lee (✉) • R. Jain • C. Ramirez
School of Engineering and Computer Science, University of the Pacific,
3601 Pacific Ave., Stockton, CA 95211, USA
e-mail: llee4@pacific.edu; rjain@pacific.edu; cramirez@u.pacific.edu

L. Stephenson
US Army ERDC-CERL, 2902 Newmark Dr., Champaign, IL 61826, USA
e-mail: larry.d.stephenson@usace.army.mil

R. Jain and L. Lee (eds.), *Fiber Reinforced Polymer (FRP) Composites for Infrastructure Applications*, Strategies for Sustainability, DOI 10.1007/978-94-007-2357-3_1,

1.1 Introduction to FRP Composites

Whereas traditional design of engineered infrastructure systems have been governed by performance and economic indicators, the increased attention towards a sustainable built environment has challenged engineers to consider environmental factors, energy/resource consumption, and social factors in addition to financial constraints and performance criteria. At the same time, fiber reinforced polymer composites (FRPs) are increasingly being considered as an enhancement to and/or substitute for infrastructure components or systems that are constructed of traditional civil engineering materials, namely concrete and steel. FRP composites are lightweight, noncorrosive, exhibit high specific strength and specific stiffness, are easily constructed, and can be tailored to satisfy performance requirements. Due to these advantageous characteristics, FRP composites have been included in new construction and rehabilitation of structures through its use as reinforcement in concrete, bridge decks, modular structures, formwork, and external reinforcement for strengthening and seismic upgrade. While the mechanical advantages of using FRP composite are widely reported in literature, uncertainty remains in regards to the feasibility of FRP composites within the framework of a sustainable-built environment.

The purpose of this book is to identify the potential of and unique challenges associated with FRP composites in a sustainable built environment. Many chapters of this book resulted from papers initially presented and discussed among participants of an international workshop sponsored by the U.S. Army Engineering Research and Development Center (ERDC), Construction Engineering Research Laboratory (CERL). The main purpose of the workshop was to gather experts from the Services, Federal Agencies, State Agencies, and academia to discuss and present ideas on the innovation of and the challenges to implementation and sustainability of FRP composites for infrastructure applications.

With consideration for the sustainability concept, this book presents recent research in two areas. The first part of the book examines various issues regarding durability and its impact on the service life of composite materials, issues that are critical in conducting life cycle assessments of composite materials in infrastructure applications. The second part of the book highlights promising applications in infrastructure systems that provide evidence for the potential of FRP composites in a sustainable built environment.

The remainder of this chapter introduces the sustainability concept and presents some of the challenges involved with the design and implementation of FRP composites in infrastructure systems. In addition, an account of specific technological challenges and applications of polymer matrix composites towards sustainable development in military installations is discussed from the perspective of an owner, specifically the Department of Defense.

1.2 Sustainability and FRP Composites

In order to evaluate the sustainability of FRP composites we must define a sustainable environment and how any material would comply within that definition. The sustainability concept, as described in the Common Future (World Commission on Environment and Development 1987) states that:

> Sustainable development is a process of change in which the exploitation of resources, the direction of investments, the orientation of technological development, and institutional change are all in harmony and enhance both current and future potential to meet human needs and aspirations. Sustainable development "...meets the needs of the present without compromising the ability of future generations to meet their own needs."

When identifying appropriate metrics for sustainability, one is likely to encounter a myriad of proposals and ideas each potentially yielding unique results; however, in general, measures are typically centered on factors that account for the following:

1. Minimum resource use
2. Low environmental impact
3. Low human and environmental health risks
4. Sustainable site design strategies
5. Higher performance

The ideal sustainable structure and material would have a closed life cycle where renewable resources, energy, and zero waste, along with minimal impact on environment and society, are considered. Certainly, there are few materials that could qualify as ideal sustainable materials and still satisfy all the performance requirements of structural systems. Even more challenging are the demands of sustainable design which essentially seeks to achieve tailored design, construction, and maintenance plans depending on impact priorities, regional issues, and economic requirements.

1.2.1 Composite Materials

A composite material is a combination of two or more distinct materials into one with the intent of suppressing undesirable properties of the constituent materials in favor of desirable properties (Astrom 1997). Over the past several decades, the use of composite materials has increased in infrastructure applications. For structural applications FRP composites are typically fabricated using a polymer matrix, such as epoxy, vinylester, or polyester and reinforced with various grades of carbon, glass, and/or aramid fibers.

The fabrication of constituent materials for FRP composites, namely matrix and fiber, are areas of concern especially when considering that the primary resources from which polymers (excluding biopolymers) are derived are crude oil, natural gas, chlorine, and nitrogen (Gerdeen et al. 2006). The most commonly used fiber reinforcements in structural applications, glass and carbon fibers, require high temperatures (1,400°C for glass; 1,200–2,400°C for carbon) during production and in some cases require petroleum as precursors. On the otherhand, even traditional civil engineering materials such as cement and steel are associated with high environmental costs (Hollaway 2010). During the manufacture of cement, intense heat, as high as 1,870°C (3,400°F), fueled by powdered coal and natural gas are used to transform raw materials into clinker. In steel, the refining process involves the use of a blast furnace using coke and limestone to heat iron ore to a temperature of 1,600°C.

When considering only energy and material resources it appears, on the surface, the argument for FRP composites in a sustainable built environment is questionable. However, such a conclusion needs to be evaluated in terms of potential advantages present in use of FRP composites related to considerations such as:

- Higher strength
- Lighter weight
- Higher performance
- Longer lasting
- Rehabilitating existing structures and extending their life
- Seismic upgrades
- Defense systems unique requirements
- Space systems
- Ocean environments

In the case of FRP composites, environmental concerns appear to be a barrier to its feasibility as a sustainable material especially when considering fossil fuel depletion, air pollution, smog, and acidification associated with its production. In addition, the ability to recycle FRP composites is limited and, unlike steel and timber, structural components cannot be reused to perform a similar function in another structure. However, evaluating the environmental impact of FRP composites in infrastructure applications, specifically through life cycle analysis, may reveal direct and indirect benefits that are more competitive than conventional materials.

1.2.2 Challenges

While literature is widely available regarding the improved performance of infrastructure materials using composite materials, there are still areas of concern that must be addressed before these materials can be considered a viable contributor towards sustainable development. The case for FRP composites as a resource for sustainable construction in nearly all applications relies primarily on the ability to develop structures that have superior service lives and will, thereby, offset the initial environmental impacts related to material processing and construction activities. In order to validate the use of FRP composites for a sustainable built environment the characterization of long-term durability and the development of life cycle cost analyses remains critical. Although FRP composites may not be an ideal sustainable material, they present a path towards sustainability that maximizes and efficiently uses natural resources from a life cycle perspective. This section will give an overview of each of the identified challenges and suggest future directions for research.

1.2.2.1 Long Term Durability

Composite materials have been successfully developed and adapted in aerospace, automotive, industrial, electrical, military, recreational, and marine industries. However, the loading conditions, environment, and manufacturing/construction techniques for civil infrastructure are significantly different with respect to long-term performance (Karbhari et al. 2003). Furthermore, the expectation for useful service life is often 50 years or more.

The expected service life and static nature of civil infrastructure presents a unique aspect to the issue of durability where environmental conditions can be singular or involve a multitude of exposures. Due to the diversity of structures and locations, a number of environments may curtail the useful life of the material and consequently the viability of FRP materials for sustainable construction. These environments can be categorized as follows (Liao et al. 1998; Karbhari et al. 2003):

- Moisture/solution
- Alkaline environment
- Thermal effects
- Fatigue
- Creep/relaxation
- UV radiation

The diverse number of applications and the harsh changing conditions characteristic of the built environment make the complete durability characterization of FRP in civil infrastructure a challenging prospect. Over the past decades, a number of researchers have engaged in the investigation of damage mechanisms and degradation behavior of FRP composites subject to a variety of exposure conditions that are singular or synergistic. A number of publications are readily available that detail the material performance of the FRP composites in the built environment. Some reviews and studies regarding durability of composites in civil infrastructure are listed here, but are by no means exhaustive (Liao et al. 1998; Karbhari 2007). As predictive models for the material properties of composites become readily available, the need to integrate available data for service life estimation and life cycle analysis will prove to be critical for accurate sustainability assessment and for comparison with other materials.

1.2.2.2 Life Cycle Assessment

The life cycle assessment of a material or structure is the sum of all the inputs and outputs in support of the item from its conception and fabrication through its operation to the end of its useful life or "cradle-to-grave" (Woodward 1997). When determining the sustainability of a material, the following items must be considered: (1) the initial costs, (2) the life of the asset, (3) the operating and maintenance costs, and (4) the disposal costs.

The initial costs of a material or structure involves capital resources, environmental and social impact resulting from a materials production and installation. While lower initial costs might be attractive in the present the uses of capital and environmental resources used and the social impact during the maintenance and replacement of a material or structure must also be addressed. It is important to note that the selection of FRP composites will not be governed by the economic, environmental, and social impacts of the material alone; but rather, the selection of FRP composites will be dependent on the contribution of the material to reducing the overall life cycle costs associated with an entire structural system.

There are five categories to consider when ascertaining the life of an asset or the selection of material (Woodward 1997):

- Functional life: period of time for which the asset is needed.
- Physical life: period of time the asset is expected to last without replacement or major rehabilitation.
- Technological life: period until it is determined that the asset is obsolete and a new superior alternative is needed.
- Economic life: period until the state of the economy dictates that a lower cost alternative is needed.
- Social and legal life: period until human desire or legal requirement dictates a replacement.

As evident from deteriorating and aging infrastructure, structures are being used far in excess of their functional life and are close to being obsolete; these structures are vital to support the economic, transportation, and societal functions. Therefore, the most attractive alternative would be to extend the physical life of the existing infrastructure or to develop new structures with far superior service lives as compared to conventional materials. Through extending the service life of a structure, a lower cost alternative is provided which is highly desirable considering the country's growing national debt. Using the five criteria state by Woodward (1997), FRP could potentially provide a highly needed, cost effective, and desirable solution to increasing the lifespan of existing assets.

1.2.3 Summary

Composite materials have developed greatly since they were first introduced. However, before composite materials can be used as an alternative to conventional materials as part of a sustainable environment a number of needs remain.

- Availability of standardized durability characterization data for FRP composite materials.

- Integration of durability data and methods for service life prediction of structural members utilizing FRP composites.
- Development of methods and techniques for materials selection based on life cycle assessments of structural components and systems.

Ultimately, in order for composites to truly be considered a viable alternative, they must be structurally and economically feasible. Numerous studies regarding the structural feasibility of composite materials are widely available in literature. However, limited studies are available on the economic and environmental feasibility of these materials from the perspective of a life cycle approach, since short term data is available or only economic costs are considered in the comparison. Additionally, the long term affects of using composite materials needs to be determined. The byproducts of the production, the sustainability of the constituent materials, and the potential to recycle composite materials needs to be assessed in order to determine of composite materials can be part of a sustainable environment.

1.3 DOD Challenges to FRP Composites and Sustainable Infrastructure

Military installations are seeking to increase their sustainability by replacing traditional steel, concrete, and wooden components with polymeric based composites. The Department of Defense (DOD) has several policies in place to promote the use of materials and processes that are renewable and environmentally sound. Two of these include E.O. 13423 (2007) and the Department of Defense Green Procurement Program. In addition, the Office of the Secretary of Defense (OSD) Corrosion Prevention and Control Program (CPC) is currently replacing, demonstrating, and implementing technologies that promote corrosion prevention and control, including the use of composite corrosion resistant materials (DOD 2010).

Composite materials have several advantages over traditionally used materials, such as steel and concrete. Among these are: (1) Very high specific strength, (2) excellent weather/water resistance, including galvanic corrosion resistance, (3) great thermal isolation properties, fire retardancy properties, and high temperature performance, (4) great freedom of shape, and (5) high degree of integration possible, which means simple integration of stiffeners, inserts, cores, and production of self supporting structures in one or two production cycles.

Some disadvantages of composite materials include the following: (1) brittleness, failing catastrophically when their strength rating is exceeded, (2) in outdoor applications, they are susceptible to damage by ultraviolet light from exposure to sunlight, and (3) polymer composites are difficult, and in some cases impossible to repair in the field. One of the chief concerns of FRP composites in military installations is the degradation of the composite due to ultraviolet (UV) radiation. UV radiation causes a photochemical effect within polymer structures, which can cause degradation of the material. The main visible effects of UV degradation are a chalky appearance and a color shift. The component surface may also become brittle. UV energy absorbed by plastics can excite photons, which then create free radicals. While many pure plastics cannot absorb UV radiation, the presence of catalyst residues and other impurities will often act as free radical receptors, and degradation occurs, and even a very small amount of impurity, e.g. trace parts per billion values of sodium in polycarbonate (PC) will initiate color instability. In the presence of oxygen, the free radicals form oxygen hydroperoxides that can break the double bonds of the backbone carbon chain leading to a brittle structure. This process is often called photo-oxidation. However, in the absence of oxygen there will still be degradation due to the cross-linking process (Arjula and Harsha 2005).

Military installations are currently either using, field testing, or considering the use of composite materials in applications where the advantages outweigh the disadvantages. Often, components are constructed by integrating polymeric composites and metallic components in order to take advantages of the beneficial properties offer by each material. These include the following applications, which have been investigated, demonstrated and implemented through research and development and demonstration programs conducted by the U. S. Army Corps of Engineers: railings, roofs, storage sheds, recycled bridges, water pipes, doors, and window frames. Selected applications are considered separately, below.

1.3.1 Railings

New ultraviolet and corrosion-resistant fiberglass reinforced polymer railings incorporate an emerging UV degradation resistant fluoroplastic layer that mitigates long-term degradation. In addition, the lightweight FRP railings can be easily replaced, if they are damaged by forklift operators. The FRP rails will have lower life cycle cost due to their lightweight standard section design, ease of implementation, inherent resistance to corrosion in corrosive atmospheres, and durability against constant bombardment by UV rays.

Even though it is corrosion resistant, normal FRP can degrade/deteriorate over time due to UV exposure. One of the main problems when considering the effect of UV rays on polymers is the intensity related to: stratospheric ozone, clouds, altitude, the position of the sun (height, time of day and time of year), and reflection. However, fiberglass rails can provide a long, maintenance-free service life in the most challenging corrosive and structural environments, when coated with a fluoroplastic protective layer such as polytetrafluoroethylene (PTFE) (Oshima et al. 2001; Shi et al. 2007). PTFE has particularly good UV resistance because it has the very strong carbon-fluorine (C–F) bond (almost 30% higher than the carbon-hydrogen (C–H) bond), which is the common side bond that surrounds the carbon (C–C) backbone in the helix and protects it. Also, most fluoropolymers do not have the light absorbing chromophore impurities in their structure that can act as an initiator for photo-oxidation.

Because of its chemical inertness, PTFE cannot be cross-linked like an elastomer. Therefore it has no "memory," and is subject to creep (also known as "cold flow" and "compression set"). This can be both good and bad. A little bit of creep allows PTFE seals to conform to mating surfaces better than almost any other plastic seal. Too much creep, however, and the seal is compromised. Compounding fillers are used to control unwanted creep, as well as to improve wear, friction, and other properties. A filler such as carbon, for example, could be added to the compound and adhered with the handrails to give them increased hardness.

1.3.2 FRP Window Frames

Fiberglass window frames are approximately three times stronger than aluminum and eight times stronger than vinyl. They can be painted and withstand weather extremes from Arctic cold to desert heat to salty sea air. In addition, fiberglass is well known as a good insulator, with R-values ranging from an insulation rating of R-7 to R-11.

Since FRP is similar to glass, the FRP frame expands and contracts at the same rate as the glass in the window which further reduces energy leaks, which may provide advantages over window frames made of wood, aluminum, or vinyl. Additionally, fiberglass takes the least amount of energy to produce and has the longest life expectancy of any window frame material, which is why leading experts say fiberglass is the most environmentally-friendly choice available.

1.3.3 FRP Roofs

Fiberglass roofs provide outstanding protection from harsh elements such as marine environments, high temperatures, temperature fluctuations, and high UV exposures. The combination of all these properties result in the delivery of a long, maintenance-free service life, even under the most demanding conditions. The resulting life cycle cost savings is very large when compared to conventional standing seam metal roofs. The new roofing materials can also be opaque in coloration, allowing diffuse light to enter the buildings, resulting in lower energy consumption when the building is in use. Installation of fiberglass roofing allows for less downtime than other systems also. Conventional UV protection additives, such as calcium carbonate, add more weight to the roof. By contrast, the fluoropolymer dispersion layer, which will be embedded in the fiberglass roofing, is lightweight and has the structural properties that make installation quick and easy.

In this case the alternative, high-durability roofing system is comprised of fiberglass polymer fabric with a fluoropolymer dispersed protective layer (such as PTFE) for additional protection against moisture uptake and UV degradation from sunlight.

Polytetrafluoroethylene has very strong molecular bonds and is highly resistant to attacks from most chemicals and UV radiation, and can retain these properties over a very wide range of temperatures, and also has an extremely low coefficient of friction. PTFE also conforms to the ASTM standards required for use in this application. The durability and corrosion resistance of this fiberglass-fluoroplastic protective layer roofing system is being compared to typical factory coated standing seam metal roof (SSMR) system in the same environment.

1.3.4 FRP Doors

For military installations in high-humidity, severely corrosive locations, preference is now given to using FRP doors and frames, with the associated hardware (hinges, screws, bolts, handles, kickplates, push-plates, closers, thresholds, panic doors, and locksets) manufactured from Type 304 stainless steel. The inner cavity of the FRP doors (filled with polyurethane foam) usually have a minimum energy efficient R-factor of 9. FRP doors are available with a flame spread rating of less than 25 according to ASTM-E84 and satisfy the self-extinguishing requirements of ASTM-D635. These doors are especially advantageous for sanitary facilities (hospitals) because the sealed, non porous outer resin is not conducive to bacteria proliferation. A number of suppliers of fiber reinforced plastic (FRP) doors for exterior use are available. At Fort Drum, NY and Fort Bragg, NC, conventional steel doors that had been severely corroded by their environments (salts to melt snow at Ft. Drum, and chemicals used in water and wastewater treatment plants at Fort Bragg) were replaced with FRP doors.

Ultraviolet degradation and mechanical strength are two limitations to be considered when specifying FRP doors and hardware. All polymers are subject to ultraviolet degradation. UV damage can be minimized by use of UV resistant additives. These additives absorb the sunlight (visible and UV) and prevent damage from proceeding further into the material. If this is done, no major damage to the component due to UV degradation should occur over a period of many years. The other limitation of FRP doors is that, while they are relatively stiff (strong) for their weight, point loading at fasteners can cause loss of contact at places where the FRP is subject to high loads. It has been found that hardware installed on FRP doors should be attached by bolting holding plates and backer plates on both sides of the doors, rather than just bolting the hardware directly to the doors, as it tends to pull out in the latter case.

1.3.5 Fiberglass Pipes

Buried fiberglass reinforced polymer tanks and pipes are cost effective alternatives to using their steel counterparts underground, thus reducing the concern of corrosion and cathodic protection, and the concomitant periodic testing requirement. Large diameter FRP piping is available up to 9 ft in diameter and designed with leak free joints, which can be accomplished with a minimum of disruption in operations. FRP pipe and joint systems are proving to be more cost effective than their concrete counterpart because of the smoother outside surface and lighter weight. These features significantly reduce the jacking pressure required and permit jacking longer runs than concrete, reducing installation costs and time.

While sewer and drainage piping continues to be dominated by concrete, there are many areas where FRP is the preferred choice. For example, concrete pipe deteriorates rapidly in sewage due to hydrogen sulfide attack. Hydrogen sulfide erodes the upper surface of the pipe and will eventually cause a cave-in. FRP is unaffected by hydrogen sulfide or purging with caustic or hypochlorite to suppress sulfide odor. As a result, FRP pipe has been used as a liner in large diameter (48–60 in.) concrete pipe.

Rehabilitation of sewer pipes can also be conducted using emerging technologies to upgrade existing piping, such as: pneumatic pulse boring through existing pipelines and horizontal drilling for the placement of new high density polyethylene (HDPE pipe) or cured in place ("sock") technologies for the in-situ application of epoxy/fiberglass material; pull-through methods to place plastic liners; and sprayed on sulfate resistant cement. The sewer line must first be cleaned and proven capable of accepting a liner. Utilizing an "impact mole," for clay pipes, the existing pipes will be pushed outward and caused to burst, clearing debris from the center and simultaneously have high density polyethylene (HDPE) pipe inserted into the void. The in-situ rehab technology can be implemented with the aid of the innovative camera inspection system to ensure that the job is correctly accomplished.

In lieu of lining with HDPE, cured-in-place-pipe (CIPP) rehabilitation may be used, especially for non-clay pipes. Using the CIPP technology, needled felt tube sock is injected with a polyester resin, and a crew feeds the sock into the existing sewer pipe using water pressure to push it inside. After the sock is in place, the water circulating through the pipe is heated to cure the resin, creating a hard, dense "pipe within a pipe," without significantly affecting the original pipe diameter. Openings for connecting lines can be cut into the new pipe once it is in place. The repair materials have been tested to meet Specification F1216 for Cured-In-Place-Pipe, which requires our materials to be tested for a minimum design life of 50 years. The CIPP technology can repair pipes made of clay, cast iron or PVC in hours without disruption to traffic or sewer flow.

Slip-lining: Slip-lining is a trench-less method of rehabilitating an existing pipe with a minimum of excavation. Centrifugal cast FRP pipe technology has advanced and yields a machine-made pipe with close outside diameter tolerances in diameters up to 96 in. The light weight and smooth outer surface permits jacking the pipe inside an existing pipe, thus rehabilitating leaking concrete sewer pipes up to 9 ft in diameter. This system of rehabilitation minimizes the jacking pressures required to push the FRP pipe through the existing concrete pipe and is done even as sewage flow continues.

Micro-tunneling: While tunnel boring has been used on large tunnel projects, micro-tunneling is a new application for trench-less piping. In micro-tunneling the FRP pipe is hydraulically jacked and pushes the cutter head through the substrata. It takes hundreds of tons of jacking pressure to push large diameter piping distances of hundreds of feet. For example, an 18 in diameter FRP pipe can be jacked at pressures up to 90 ton and 9 ft diameter FRP pipe at pressures up to 1,750 ton. In the past, stainless steel sleeves have been used as a reinforcement around concrete pipe joints to withstand hydraulic jacking pressures. However, FRP sleeve joints are proving to be a cost-effective replacement for stainless steel used in concrete pipe jacking.

Fire Protection: For fire protection deluge systems, FRP piping involves FRP fire resistant material systems to replace the steel pipes, including sprinkler systems. Steel is susceptible to corrosion, causing steel fire protection water conveyance pipes to leak, or nozzles and sprinkler heads to plug up with corrosion products. FRP systems do not suffer from this disadvantage.

1.3.6 Bridges from Recycled Plastic

The DoD is interested in recycled plastic as a possible replacement for wood-timber bridges, which are very costly to repair, often exceeding the cost of replacement, at all military installations. Also, if the same materials are used, the same degradation cycle begins all over again. A replacement for a wooden bridge made of ~85,000 lb of recycled co-mingled polymers, including high density polyethylene, glass, and vehicle bumpers, was recently installed at Ft. Bragg, NC (Larsen 2010). The composite components included I-beams railings, decks – only the fasteners were made of corrosion resistant steels. The bridge has a 36 ft span, and was designed to support a 71 ton load.

The goal was to provide a low-maintenance, affordable structure using recycled materials and avoiding the use of any wood components which require chemical treatments to fight rot and insect attack, as well as costly routine maintenance to repair or replace deteriorated members.

The Office of the Secretary of Defense (OSD) Corrosion Prevention and Control Program (CPC) (DOD 2010) funded the plastic bridge's design and the Army paid for its construction and initial load testing. CPC also is funding continuous, remote monitoring. For the load testing, the Army Bridge Inspection Team, along with their contractor recorded measurements from various strain and deflection gauges mounted on the bridge. To complete the load testing, different types of vehicles, in addition to an M-1 Abrams Tank, crossed the bridge multiple times. In addition to assessing the bridge's structural performance under heavy loading, the bridge is being equipped with special sensors and video cameras to remotely monitor the durability of the thermoplastic materials for long-term use in such a bridge. Fort Bragg has built a second thermoplastic composite bridge and has approval to construct a third. These structures will also be monitored to further validate the technology and provide lessons learned.

1.4 Environmental Implications of FRP Composites

While the advantages of FRP composites with respect to conventional materials are well-known throughout literature, the synthetic nature of the constituent materials used to manufacture FRP composites presents uncertainty and concern regarding the materials' environmental impacts from processing, construction application, in-service operation, and disposal at the end of life. The specific environmental impacts include, but are not limited to, concerns over health, safety, emission of volatile organic compounds, energy consumption, and toxicity. Certainly the use of crude oil to derive constituent materials of a composite, namely fiber and matrix, present considerable doubt with respect to the sustainability of FRP composites; however, when examined over an infrastructure component or system's entire life cycle, the use of FRP composites may potentially reduce the carbon footprint and require less energy when integrated with conventional materials.

The impact of materials selection and manufacturing on the environment has garnered the attention of design engineers. The first step in minimizing pollution and environmental damage due to manufacturing processes and materials requires characterizing the sources of pollution and environmental damage during the life cycle of a system or component. The manufacturing and construction processes are associated with the use of raw materials, energy, and water resources, and the generation of emissions to air, water, and land. For FRP composites, the use of natural resources and emissions must be considered as part of the manufacturing sequence, which involves material selection, mold preparation, preparation of materials, application of materials, mold operations and consolidation, curing and post curing, trimming the component, and, finally, cleaning and disposal of equipment.

Conducting an accurate assessment of environmental impact requires knowledge of the component or system and the construction processes involved with all materials as part of a life cycle analysis from raw materials input, system maintenance, and decommission. However, the information needed to fully characterize the environmental impacts of fiber and matrix choices is limited. For composite materials, the ability to alter component or system performance by tailoring composite characteristics (i.e. constituent material properties, volume fractions, orientation, etc.) is one of primary draws of the material to the design engineer, but it is this same tailorability which presents a challenge to life cycle analysis. The challenge is that each constituent material has a complex set of inputs and outputs that result in its existence (Calkins 2009).

A sustainability assessment typically involves science-based weighting of the environmental, social, and economic impacts or consequences of actions during a product's life cycle from cradle to grave (Singh et al. 2009). The remainder of this chapter highlights areas associated with the manufacturing of constituent materials and methods of processing of composite materials that are prevalent for civil infrastructure and present the most significant sources of environmental impact of FRP composites. It is understood that only through a comparative life cycle analysis from cradle-to-grave of a specific product that a judgment on the superiority of one material selection over another can be made. The intent here is to draw attention to the manufacturing and processing aspects of composite materials in civil infrastructure applications that are likely to result in environmental impacts.

1.4.1 Environmental Impacts of Constituent Materials

Composites materials have been identified as a versatile material for its wide variety of applications, satisfaction of performance requirements, and, in some cases, lowering overall project costs. In order to evaluate the sustainability of FRP composites, the constituent materials, namely fiber and matrix, and manufacturing techniques must be considered. The most common polymers used as composite matrices are polyesters, epoxies, and vinylesters. Due to the use of fossil fuels, polymers have considerable environmental impact associated with raw materials extraction and are often the primary cause of environmental impact, as compared to the fibers, of the FRP composite material (Anderson et al. 2004). The following sections provide an overview of the most commonly used polymers and fibers in infrastructure applications and discuss associated challenges towards sustainability.

1.4.1.1 Epoxy

Epoxy resins are preferred in civil infrastructure applications because of their high mechanical strength, low viscosity, low shrinkage rates, and environmental durability; however, they are relatively expensive when compared to other polymers (Kaw 2005; Bank 2006). An epoxy system typically involves an epoxy resin and a hardener which are supplied in two parts and are mixed in specific proportions to cause the curing reaction (Bank 2006). A variety of hardeners and accelerators are employed to promote crosslinking between molecules of an epoxy. Some epoxies can crosslink at room temperatures, but most epoxies used in composite applications require an increased temperature to initiate the crosslinking, and well controlled temperature throughout the process to crosslink as intended (Astrom 1997).

In general, the toxicity of an epoxy decreases with increasing molecular weight (Astrom 1997). Most hardeners have high levels of toxicity and are severe irritants and sensitizers when touched or inhaled (Astrom 1997). Some hardeners are severely corrosive, some are known carcinogens and others can cause damage to liver and kidneys (Astrom 1997). Additionally, since epoxy is a thermoset, the curing process is irreversible. While a number of researchers have investigated recycling of epoxy resins, it remains a challenging procedure with limited applications (Dang et al. 2002).

1.4.1.2 Polyester

Polyesters are advantageous because of their low cost, high corrosion resistance, and uncomplicated processing. They can also be applied to a variety of applications.

The main disadvantages are low temperature tolerance, shrinkage from crosslinking, and potential health problems. The polymerization reaction for polyester usually takes place in the presence of a catalyst, such as peroxide (Bank 2006). Peroxide is toxic and can irritate eyes and skin. Styrene, a crosslinking agent, poses health concerns, and its emissions are a "significant source of impact during both the mixing and application of a resin" (Anderson et al. 2004). Styrene easily evaporates, so it can be inhaled, which can irritate eyes and the respiratory tract. It can also lead to more serious effects, such as brain damage and changes in blood cells (Astrom 1997).

In a comparison of polymer matrices used to fabricate a 1 m^2 double curvature panel, it was found that polyester had the highest environmental impact, while epoxy resin had the second most impact (Anderson et al. 2004). It should be noted that the used of fillers, such as calcium carbonate, reduced the environmental impact of polyester by half since the volume of resin needed was reduced.

1.4.1.3 Vinylester

Vinylester combines the fast and simple crosslinking of unsaturated polyesters with the mechanical and thermal properties of epoxies. Its corrosion resistance and durability in alkaline environments have made vinylester resin a popular choice in infrastructure applications (Bank 2006). The disadvantage of vinylester is that there are fewer crosslink sites per molecule. Vinylester is produced from an epoxy and acrylic ester monomer. The crosslink times for vinylester molecules are similar to unsaturated polyesters and can occur at room temperature or elevated temperatures (Astrom 1997). Like epoxy, vinylester is also a thermoset, making it very difficult to recycle at the end of its lifespan.

Vinylester is produced in a combination of the method for polyester production and the method for epoxy production. Styrene and a peroxide catalyst are common contributors to production, just as in polyester (Bank 2006). Therefore, some of the same emissions and health effects are possible with the production of vinylesters.

1.4.1.4 Carbon Fiber

Carbon fibers are manufactured from three different constituents: rayon, polyacrylonitrile (PAN), and petroleum pitch. Fibers are initially drawn and oxidized at temperatures below 400°C. After this process, the fibers undergo pyrolysis (heated above 800°C). Finally, graphitization is carried out. The fibers are heated to temperatures above 1,000°C.

The advantageous properties of carbon fibers include high tensile strength, high modulus of elasticity, and a high tolerance to temperature and corrosive environments; marine environments are one of the major concerns in civil infrastructure so materials resistant to corrosion are desirable. However, carbon fiber is petroleum-based. Furthermore, carbon fiber requires high temperatures to manufacture, requiring the use of a furnace. Carbon fibers have the greatest environmental impact out of the three types used for structural applications. This is because of the high energy demand of the manufacture of carbon fiber. However, because carbon has high performance characteristics, a lower quantity of material is often needed to achieve the same performance for a longer duration than glass or aramid fibers potentially offsetting the environmental impact (Anderson et al. 2004).

1.4.1.5 Glass Fiber

Two commonly used glass fibers in the industry are E-glass and S-glass. The primary material in most glass fibers is silica; to effectively produce theses fibers, the ingredients must be melted in a furnace that at temperatures of about 1,370°C.

Glass fibers are a popular choice for fiber reinforcement due to advantageous properties such as high strength (tensile strength of approximately 3.40 GPa), tolerance to high temperatures and corrosive environments, and low cost. However, glass fibers have a relatively low stiffness. Typically values of stiffness for glass fibers range from approximately 70 to 90 GPa, whereas, the stiffness of carbon fiber can range from 230 to 830 GPa. Additionally, glass fibers are also very sensitive to moisture; in a study conducted by Abdel-Magid (2005) the strength of a glass fiber specimen decreased by 35% (767–499 MPa) when exposed to water at room temperature while under a constant tensile load (Astrom 1997). This is a concern if glass fibers are to be considered for areas that are particularly humid.

Glass fiber production uses much energy and fossil fuels. To produce a glass fiber mat, much nonrenewable energy is required. A total of 54.7 MJ/kg of nonrenewable energy are used (Joshi et al. 2004). This includes energy used for each step, including: raw materials, mixture, transport, melting, spinning, and mat production. In addition, glass processing is dusty, which makes the production of glass fibers a contributor to air pollution (Daniel and Nagtegaal 2001). Like carbon fibers, glass fibers require high temperatures to manufacture. In order to assess the sustainability of glass fibers, the amount of carbon emission and byproducts produced as a result of manufacturing need to be quantified for carbon fiber and all other basic construction materials.

1.4.2 Environmental Impacts of Production

It has been emphasized throughout this chapter that an adequate comparison of environmental impacts due to material choice requires a life cycle analysis that depends on the component or system being constructed and its expected service life. Each material choice is characterized by its respective production stages. Figure 1.1 illustrates a typical production process for a component or system constructed of FRP composites within the framework of life cycle assessment. Most composite manufacturing processes require the use of energy, raw materials, and water resources and can result in various forms of emissions to air, land, and water. Each stage of the manufacturing process has an environmental impact that contributes to the total impact of the composite component; there are multiple options in each

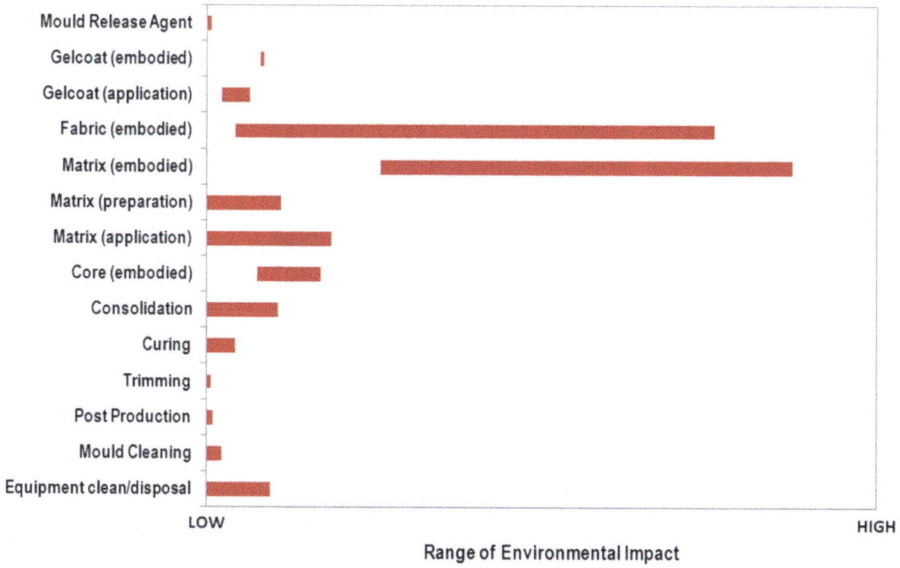

Fig. 1.1 Overview of composite manufacturing process (Anderson et al. 2004)

manufacturing stage which results in varying magnitude and range of environmental impact (Anderson et al. 2004). Figure 1.2 provides a qualitative chart illustrating the varying range and magnitude of environmental impacts for each step in the manufacturing process.

While the figures provide a qualitative guide to materials selection for composite materials from a life cycle context, the need exists for quantitative analyses regarding input resources and emissions for all materials for a given product in order to maximize the sustainability of civil infrastructure components and systems. Some researchers have initiated the process of comparing the economic, environmental, and social costs of constructing civil infrastructure systems with FRP composites versus conventional materials such as steel and concrete (Halliwell 2010; Daniel and Nagtegaal 2001). For instance, Daniel and Nagtegaal (2001) compared initial construction costs, maintenance costs, embodied energy, and environmental impact for the replacement of a corroded steel pedestrian bridge. Five materials (structural steel, stainless steel, glass FRP composites, aluminum, and concrete) were analyzed and compared to find the material that offered the most sustainable solution. Tables 1.1–1.3 show the comparison of environmental impact, emissions to air, and emissions to water, respectively, of all the materials. It was found that for combined initial construction and maintenance costs that structural steel was the least expensive followed by FRP composites, aluminum, and stainless steel. In terms of embodied energy, the FRP composite consumed the least amount of energy at 120,000 MJ, while all other material options consumed at least twice as much. The FRP composite was also found to have the least environmental impact in terms of emissions to air and water.

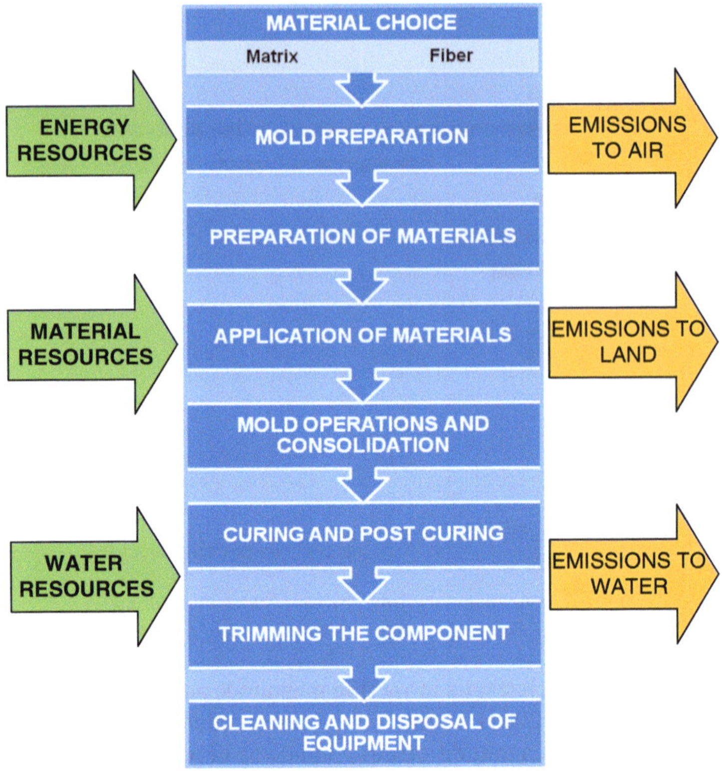

Fig. 1.2 Environmental impacts of composite manufacturing processes (Anderson et al. 2004)

Table 1.1 Comparison of environmental impact of five structural materials (Daniel and Nagtegaal 2001)

	Criterion	
Material of the bridge	Energy consumption (MJ)	Critical volume of polluted resource
Structural steel	294,000	Water: 697.4 m^3 Air: 7,090,000 m^3
Stainless steel	329,600	Not investigated, but estimated to contribute more water and air pollution than structural steel
Composites	120,000	Water: 85.8 m^3 Air: 7,920,000 m^3
Aluminum	268,700	Water: 565.3 m^3 Air: 41,100,000 m^3
Concrete	277,200	Water: 341.9 m^3 Air: 31,040,000 m^3

Table 1.2 Comparison of emissions to air of structural materials with threshold (Daniel and Nagtegaal 2001)

	Structural steel	FGRP	Aluminum	Concrete	Threshold
Polluter	kg/m^3 of product	kg/m^3 of product	kg/m^3 of product	kg/m^3 of product	kg/m^3 of water
CO_2	2.56E+03	1.03E+03	2.10E+04	4.95E+02	9.00E–03
CO	9.58E+01	1.32	5.15E+01	3.48	4.00E–05
CH_4	5.95	1.21	5.39E+01	9.89E–01	6.70E–03
N_2O	3.70E–02	4.80E–03	2.94E–01	1.51E–02	1.00E–07
dust Fe/Al oxi	2.20E–01	1.05E–01	1.65	6.00E–02	1.00E–07
dust Si/CA oxi	4.20E–02	5.05E–01	2.70E–01	4.70E–01	3.00E–07
SO_2	3.28	2.51E–03	1.27E+01	2.80E–01	1.20E–06
NO_x	3.08	2.83	2.45E+01	1.27	1.00E–05
Styrene	–	1.20E–01	–	–	8.00E–07

Table 1.3 Comparison of emissions to water of structural materials with threshold (Daniel and Nagtegaal 2001)

	Structural steel	FGRP	Aluminum	Concrete	Threshold
Polluter	kg/m^3 of product	kg/m^3 of product	kg/m^3 of product	kg/m^3 of product	kg/m^3 of water
Aluminum	3.33E–06	2.00E–06	3.09E–05	1.65E–07	5.00E–05
Ammonia	4.58E–03	1.10E–03	4.23E–02	2.38E–04	2.20E–03
Cadmium	4.57E–05	2.10E–06	4.28E–04	2.18E–06	3.50E–06
Copper	1.96E–08	7.90E–04	1.82E–07	9.90E–10	2.00E–04
Cyanide	3.08E–04	7.40E–05	2.85E–03	1.60E–05	1.00E–04
Fluoride	1.03E–01	2.00E–04	6.49E–03	3.51E–03	1.50E–03
Manganese	6.07E–06	3.60E–06	5.64E–05	3.03E–07	5.00E–05
Mercury	1.57E–04	7.00E–07	1.45E–03	7.53E–06	5.00E–06
Zinc	3.97	1.40E–03	5.44E–02	1.35E–01	5.00E–03
Cobalt	–	3.00E–02	–	–	1.00E–03

The study by Daniel and Nagtegaal (2001) provides evidence for the competitiveness of FRP composites in a specific application. In general, materials that compete with composites are more susceptible to degradation. Degradation can lead to more maintenance or the need for the material to be replaced often. This means that composites impact the environment less than their competitors because they require less energy use due to the lower need for replacement. FRP bridge decks, for example, have three times greater service life, on average, than that of concrete (Hollaway 2010). Therefore, less waste is generated from FRP composites because replacement is necessary after a longer duration of time.

The low density of FRP composites help facilitate installation by minimizing the need for heavy equipment. In addition, FRP composite components for structural systems are manufactured in properties as opposed to cast-in-place. This allows for the quality of the composite to be monitored, unpleasant weather to be avoided, and light lifting cranes to be required. Other benefits of heavy equipment not being required are that much energy, time, and costs are saved. (Hollaway 2010). Less labor (included in cost) and traffic control are needed than for using competitive materials. Also, fewer emissions are released to the environment because the equipment will not be used for long durations, if used at all.

1.4.3 Summary

The advantageous properties of FRP composites are well understood and evidence exists that FRP composites can be a viable materials selection toward a sustainable built infrastructure when considering direct and indirect benefits throughout a component or systems life cycle. In addition there is evidence that FRP materials can offer several environmental advantages, namely (Halliwell 2010): (1) as long as the FRP composites exists no fossil derived CO2 is admitted; (2) little water is needed during manufacturing; and (3) the lightweight of the material minimizes the need for heavy equipment and thus minimizing fuel consumption and emissions during construction. However, before composite materials can be used as an alternative to conventional materials as part of a sustainable environment a number of needs remain.

- Development and availability of standardized energy consumption and emissions data for FRP composite constituent materials.
- Integration of durability data and methods for service life prediction of structural members utilizing FRP composites.
- Determine appropriate strategies for end of life to minimize waste.
- Development of formalized methods and techniques for life cycle assessment of structural components and systems.

Ultimately, in order for composites to truly be considered a viable alternative, they must be structurally and economically feasible. Currently, there are many studies regarding the structural feasibility of composite materials. Very little quantitative research has been done on the economic and environmental feasibility of these materials from the perspective of a life cycle approach.

References

Anderson J, Jansz A, Steele K, Thistlethwaite P, Bishop G, Black A (2004) Green guide to composites: an environmental profiling system for composite materials and products. BRE Bookshop, Watford, Retrieved from www.brebookshop.com

Arjula S, Harsha AP (2005) Study of erosion efficiency of polymers and polymer composites. Department of Mechanical Engineering, Institute of Technology, Banaras Hindu University, Varanasi. Dec 2005

Astrom BT (1997) Manufacturing of polymer composites. Chapman & Hall, New York

Bank LC (2006) Composites for construction: structural design with FRP materials. Wiley, Hoboken

Calkins M (2009) Materials for sustainable sites. Wiley, Hoboken

Daniel RA, Nagtegaal G (2001) Pedestrian bridge of pultruded sections as result of ecological design. In: Proceedings of the EPTA seminar, Roermond, Oct 2001

Dang W, Kubouchi M, Yamamoto S, Sembokuya H, Tsuda K (2002) An approach to chemical recycling of epoxy resin cured with amine using nitric acid. Polymer 43(10):2953–2958.

Department of Defense (2010) INSTRUCTION NUMBER 5000.67, USD(AT&L), SUBJECT: prevention and mitigation of corrosion on DoD military equipment and infrastructure, 1 Feb 1 2010, Washington, DC

Executive Order 13423 (2007) Strengthening Federal Environmental, Energy, and Transportation Management. http://www.archives.gov/federal-register/executive-orders/2007.html

Gerdeen JC, Lord HW, Rorrer RAL (2006) Engineering design with polymers and composites. CRC Press, Boca Raton

Halliwell S (2010) FRPs – the environmental agenda. Adv Struct Eng 13(5):783–791

Hollaway LC (2010) A review of the present and future utilization of FRP composites in the civil infrastructure with reference to their important in-service properties. Constr Build Mater 24:2419–2445

Joshi SV, Drzal LT, Mohanty AK, Arora S (2004) Are natural fiber composites environmentally superior to glass fiber. Composites Part A 35:371–376

Karbhari V (ed) (2007) Durability of composites for civil structural applications. Woodhead Publishing Limited, Cambridge/England

Karbhari VM, Chin JW, Hunston D, Benmokrane B, Juska T, Morgan R, Lesko JJ, Sorathia U, Reynaud D (2003) Durability gap analysis for fiber-reinforced polymer composites in civil infrastructure. J Compos Constr 7(3):238–247

Kaw AK (2005) Mechanics of composite materials, 2nd edn. CRC Press, Boca Raton

Liao K, Schultheisz CR, Hunston DL, Brinson LC (1998) Long-term durability of fiber-reinforced polymer-matrix composite materials for infrastructure applications: a review. J Adv Mater 30(4):3–40

Larsen KR (2010) Corrosion-resistant thermoplastic is constructed of recycled milk jugs and automobile bumpers. Mater Perform 49(1):22–27

Oshima A, Udagawa A, Tanaka T (2001) Fabrication of polytetrafluoroethylene/carbon fiber composites using radiation crosslinking. Radiat Phys Chem 62(1):77–81

Shi YJ, Feng X, Wang HY, Liu C, Lu XH (2007) Effects of filler crystal structure and shape on the tribological properties of PTFE composites. Tribol Int 40(7):1195–1203

Singh RK, Murty HR, Gupta SK, Dikshit AK (2009) An overview of sustainability assessment methodologies. Ecological Indicators 9:189–212

Woodward DG (1997) Life cycle costing – theory, information acquisition and application. Int J Project Manag 15(6):335–344

World Commission on Environment and Development (1987) Our common future. Oxford University Press, Oxford

Chapter 2
Sustainability in Infrastructure Design

Hector Estrada, Delicia H. Borja, and Luke Lee

Abstract It is well understood that the main objective of infrastructure design code specifications is to protect the public's welfare, health, and safety; none of which appear to be directly related to sustainability, which as a movement focuses attention on protecting the natural environment, conserving resources, and minimizing the toxicity of construction materials and processes. However, a number of jurisdictions have adopted language based on the United States Green Building Council to curtail the adverse effects of global warming gases and minimize environmental impact of new construction; in some cases to improve air quality in the community and to increase the long-term viability of local construction. In this chapter, we provide a survey of the available literature and review recent, and some pending, changes to design specifications found in building codes related to sustainability; particularly related to structural concrete, steel, timber, and other construction materials. These materials' code writing entities have made a commitment to provide guidance to the design professionals regarding sustainability. Some of the groups are further along than others, but all have working groups (as of early 2011). Also, the International Code Council, Inc. has written a public version of the International Green Construction Code with the final version slated to be published in 2012; this is the first step towards making sustainability common place in all infrastructure design.

H. Estrada (✉)
Department of Civil Engineering, University of the Pacific, 3601 Pacific Ave., Stockton, CA 95211, USA
e-mail: hestrada@pacific.edu

D.H. Borja • L. Lee
School of Engineering and Computer Science, University of the Pacific, 3601 Pacific Ave., Stockton, CA 95211, USA
e-mail: d_borja@u.pacific.edu; llee4@pacific.edu

R. Jain and L. Lee (eds.), *Fiber Reinforced Polymer (FRP) Composites for Infrastructure Applications*, Strategies for Sustainability, DOI 10.1007/978-94-007-2357-3_2,

Keywords Sustainability • LEED • Steel • Concrete • Timber • Design specifications • Performance-based structural design

2.1 Introduction

With recent changes in governmental policies to address, in part, increased awareness of concerns over greenhouse effects on the environment and more effectively utilize our limited natural resources, sustainability has become an important aspect in infrastructure design. Current innovations in sustainability are spearheaded by professional organizations such as the American Society of Civil Engineers (ASCE), the American Institute of Steel Construction (AISC), the American Concrete Institute (ACI), the American Forest & Paper Association (AF&PA), and the U.S. Green Building Council (USGBC). Through extensive work from the aforementioned organizations, various changes in standards and specifications have been developed, proposed to building code officials, and some have already been implemented.

These professional organizations' main goals, from a sustainability standpoint, are to reduce the carbon footprint of structures, curtail any negative effects on the environment caused by the construction of buildings, and improve indoor environmental quality (particularly indoor air quality). A further goal of these professional organizations is to reduce social and economic impact through the use of a sustainability certification of existing and new structures to rate their overall impact on the environment, society, and the economy. The main path for attaining all these goals is by working with engineers and architects to implement changes that reflect sustainability methods in building codes; with the U.S. Green Building Council leading the way through their Leadership in Energy and Environmental Design (LEED) certification program, which serves to standardize and rate the overall sustainability of a building using a point system.

The LEED certification program has created standards of design and development for new construction, existing buildings, commercial buildings, residential homes and neighborhoods. It awards points based on site selection (Site Sustainability, Location and Planning), Water and Energy Efficiency, Materials and Resources Used, Indoor Environmental Quality (including indoor air quality), Innovation in Design, and Regional Priority. For new construction ratings in LEED v3 (version 3 is the current rating), buildings are awarded points on a 100-point scale (plus an additional 6 points for Innovation in Design and 4 points for Regional Priority) and classified as certified (40–49 points), Silver (50–59 points), Gold (60–79 points), or Platinum (80 points and above).

Of the eight major areas that LEED addresses, Materials and Resources and Innovation and Design are directly related to sustainable materials, which is the main theme of this book. Also, two other areas are indirectly related to sustainable materials: Energy & Atmosphere and Indoor Environmental Quality; particularly when considering that a great deal of the carbon emissions and other pollutants are generated in the mining or harvesting, processing, transportation, and installation of construction materials; discussed later in the Pollution section. Therefore, a high level of sustainability cannot be achieved without detailed consideration of materials in infrastructure design. Various methods to address sustainability during the design and construction phases include specifying recycled materials or materials from sustainably managed sources, reusing parts of demolished structures (deconstruction), specifying locally available materials, using prefabricated components whenever possible, and, more importantly, designing structures that can be altered and can be adapted to new uses (reuse) or loading conditions.

Although building sustainable infrastructure has historically been seen as more expensive and less profitable for developers, recent advances in materials, processes, and equipment make the cost of building "green" competitive. Also, changes in industry standards and codes have enabled sustainability to become widely accepted in building design and construction. Some of the most important changes have been spearheaded by government agencies through the adoption of LEED certification for public projects; for example, requiring all public governmental buildings to be LEED certified, including public academic institutions. Although there is a 1–2% increase in building construction costs associated with achieving the lowest LEED certification (Kats 2003), the lifecycle cost (including energy consumption of the building) is lower as compared to the costs of conventional construction (this will be discussed further in the next section).

To increase public sector investment in development using environmentally friendly techniques, particularly LEED certified structures, there are a number of incentives from governmental policies, including grants, tax credits, and low interest loans. Furthermore, substantial benefits can be achieved due to the increased productivity resulting from better ventilation, temperature control, lighting control, and reduced indoor air pollution; all also contributing to better living spaces. In fact, several studies since 2008 have found that LEED for-rent office spaces generally achieve significantly higher rents, sale prices, and occupancy rates (Miller et al. 2008). All of these incentives should present lower investment risks and much higher profits as compared to conventional construction.

2.2 Cost of Sustainable Infrastructure Materials

As the cost of construction materials continues to increase because of the ever increasing costs associated with production, there has been a move toward use of renewable and more sustainable materials in infrastructure design. In recent years, there have been changes to concrete, steel, wood, and masonry productions in efforts to provide consumers a more sustainable product while reducing economic and social adverse impacts. Even though more cost effective production methods are currently in use, there is still an average increase of approximately 2% in the overall cost associated with "green" construction when compared to conventional building construction (see Table 2.1). Table 2.1 lists the percent initial cost premium associated with each of the four LEED rating levels (Kats 2003). The table includes data

Table 2.1 Average cost increase associated with different levels of LEED rating (Kats 2003)

LEED rating	Number of projects	Green cost premium
Certified	8	0.66%
Silver	18	2.11%
Gold	6	1.82%
Platinum	1	6.50%
	Total = 33	Average = 1.84%

from 33 LEED registered projects; 25 are office buildings and 8 are schools. It is interesting to note that the average increase for gold-LEED certification is lower than that for silver-LEED certification. There are two primary reasons, there are more LEED-silver projects than LEED-gold and the LEED-gold projects were constructed more recently; this later reason is a result of "economies of scale" – more LEED certified buildings lower cost for sustainable construction/equipment (derived partially from learning by doing). Also, the only platinum building in the data had an increase of 6.5% in cost. Note that this initial cost premium for high-performance can be offset over the life of the building by the optimized design and project quality components. This is the conclusion reached by the U. S. Department of Energy; that is, the average 2% initial cost associated with construction of sustainable high-performance structures is rather small when compared to the lifecycle cost savings associated with a more efficient structure (Kats 2003).

The long-term cost savings for high performing buildings can be substantial. A study of 60 LEED certified buildings in 2003 concluded that these buildings were on average 25–30% more energy efficient (Kats 2003). And given that over 40% of a building's energy consumption (not including the energy required to manufacture all building materials and construction) is associated with its operation, a 25% savings can result in as much as a 10% reduction in a building's energy consumption. This translates in over a 10% return on the initial investment – just on operation alone. It is clear that choosing a sustainable design with a modest increase in initial cost can reduce operation costs during the building's lifetime, resulting in substantial savings. The potential lower lifecycle cost is the primary reason why a number of private and public entities are specifying LEED certified structures.

In a typical building, the structural system usually accounts for 10–20% of the construction costs (depending on building use), with office buildings having the greatest percentage (18%) of construction cost devoted to structural components (Kneer and Maclise 2008). Therefore, the increased cost associated with the structural materials for LEED compared to non-LEED construction is only between 0.2% and 1.2% (i.e., 18% of the average increase of overall costs associated with LEED compared to non-LEED construction as listed in Table 2.1).

An additional cost of sustainability is the emerging risks of potential legal issues with sustainable construction. Bristol and Dolan (2008) warn that the demand for sustainable design and construction may outpace the acquisition of knowledge and skills of most engineers and others in the construction industry. Of particular concern is litigation resulting from expanded warranties, breach of contract, or issues from untested "green" products and technologies that have been hastily used to achieve a specific LEED rating level. Brinson and Dolan (2008) recommend that engineers address these risks upfront in the contracts by explicitly stating the duties and responsibilities of deign and construction members, and by clearly defining the expected performance of "green" products.

2.3 Pollution Impact of Infrastructure Materials

As noted earlier, two sustainability areas (Energy & Atmosphere and Indoor Environmental Quality) are indirectly related to sustainable materials. These areas are primarily concerned with control of pollutants and their effects on public health and the environment; particularly curtailing carbon dioxide (CO_2) emissions to reduce climate change, and control other pollutants that adversely affect human health. Considering that only approximately 20% of a building's environmental impact is embodied in the materials (the other 80% is contributed by operations), it would appear that the environmental effect of construction materials is minimal (Hays and Cocke 2009). However, all of the environmental impact from construction materials happens initially, before operations have any environmental impact, which happens over the life of buildings. Noting that approximately 37% of CO_2 released today will be present in the atmosphere 100 years later (Kang and Kren 2007); and considering the conclusions of the Intergovernmental Panel on Climate Change regarding the catastrophic changes to the environment expected if drastic steps are not taken immediately to reduce CO_2 levels in the atmosphere (Mehta and Meryman 2009), global warming effect from materials should be expected to be much more severe than those of operations (with more time society will be able to develop more carbon-neutral operations systems). This has been one of the primary factors why manufacturers (primarily motivated by consumers) have made substantial changes toward using more sustainable practices that have a lesser, immediate impact on the environment.

Also, current laws have incentivized limiting the amount of pollutants produced in the manufacturing of concrete and steel. Therefore, concrete and steel production companies have opted to reduce energy consumption and recycle concrete and steel in order to comply with current environmental protection laws and reduce processing costs. Limitations of air emissions and waste from manufacturing have both contributed to environmental improvements in both the concrete and steel industries.

The carbon footprint from cement (the main ingredient in concrete) had nearly doubled between 1990 and 2005; Fig. 2.1 depicts historic and projected annual cement consumption rates. If annual cement consumption rates continue along their current trend, we can expect pollution levels from cement production to double about every 5 years. To produce cement, calcium carbonate ($CaCO_3$) must be converted into calcium oxide (CaO) as follows (Mehta and Meryman 2009):

$$CaCO_3 + \text{Heat} \rightarrow CaO + CO_2$$

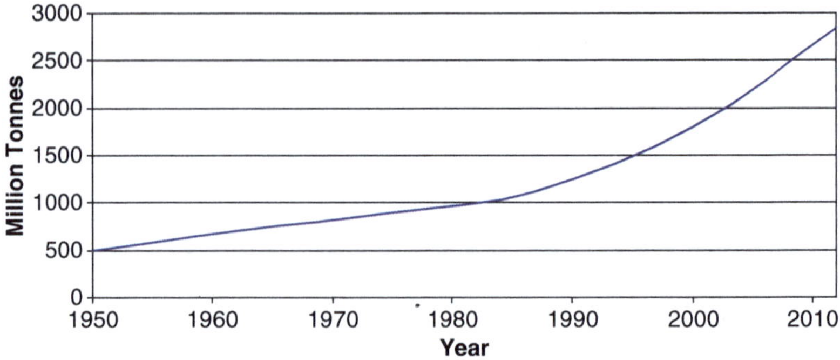

Fig. 2.1 Historic and projected annual cement consumption rates (Mehta and Meryman 2009)

Table 2.2 Portland Cement Association recommendations for fly ash and slag cement replacement (Kang and Kren 2007)

Product	PCA/ACI recommended replacement by weight	Comments
Fly Ash Class C	15–40%	Pozzolanic and cementitious properties Generally available on the east coast and the midwest Not allowed in California hospitals or public schools [California Building Code]
Fly Ash Class F	15–25%	Pozzolanic properties Generally available on the west coast, allowed in California hospitals and public schools [California Building Code]
Slag	30–40% Typical	

The main by-product of the chemical reaction and that of the combustion of fossil fuels to generate the Heat is CO_2. It is estimated that a ton of CO_2 is produced to manufacture a ton of cement. When the cement is combined with aggregate and water to form concrete, this figure goes down to about a third of a ton of CO_2 per ton of concrete (Falk 2010). Therefore, to reduce the carbon emissions associated with concrete, we must reduce the amount of cement used in concrete. To accomplish this, cement manufacturers have recently resorted to adding fly ash or slag into cement mixtures. For example, Europe uses Portland-slag cements containing over 50% ground granulated-blast-furnace slag (Mehta and Meryman 2009) as cement replacement in concrete. Also, the Portland Cement Association has developed recommendations for fly ash and slag cement replacement by weight, as listed in Table 2.2. PCA also lists the replacement allowed depending on the region where the construction project will be located.

Production of cement and concrete also generates large amounts of dust that can escape into the atmosphere. In order to regulate this, the Environmental Protection

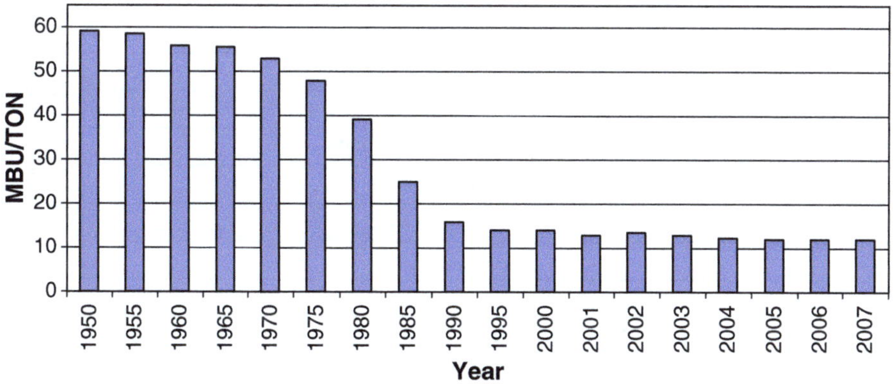

Fig. 2.2 Energy consumption per ton of steel shipped in the U.S (Stubbles 2000)

Agency has placed a limit on the quantity of dust that is released into the atmosphere. Therefore, cement manufacturers are recycling the dust emitted during manufacturing, which can be done with mechanical collectors, electrical precipitators, or fabric filters. Recycling the dust is a complex process because of its high alkalinity, which must be reduced before it is used. This recycled material can be used in agricultural soil treatment, diverting it from landfills. Sulfur emissions are another concern along with CO_2 that results from cement and concrete production.

As for steel, because of its high recyclability, a large amount of deconstructed and scrap steel is used to manufacture new steel. This greatly decreases the energy consumption needed to produce a ton of steel. In fact, as shown in Fig. 2.2, the energy consumption per ton of steel shipped in the United States, on average, has decreased by about 75% over the past half century (Stubbles 2000). The most drastic change came in the 1970s and 1980s when the industry began using recycled steel and switching from coal burning furnaces to electricity. The carbon footprint from steel is relatively small (0.73 ton of CO_2 are produced to manufacture a ton of steel) considering its high strength-to-weight ratio (AISC 2011). Also, as shown in Fig. 2.3, the iron and steel sector's normalized air toxicity releases show a 70% decline from 1994 to 2003 and appear to have leveled out. Most (99%) of pollution releases are air, with less than 1% water release. The main pollutants released into the air are manganese, chromium, and lead (US EPA 2006). Also, if current trends in recycling and reusing of structural steel continue, steel may become a carbon-neutral material. Furthermore, Table 2.3 lists the total greenhouse emissions in teragrams of CO_2 equivalents from 1990 and 2005. For iron and steel production, there was a decrease of about 47% in the CO_2 emissions. The incorporation of recycled steel in the production of new steel has drastically decreased the carbon footprint of the industry.

Other energy conservation measures adopted by the steel industry include consolidation of the industry at product plants that are more modern (know as mini-mills) and process improvements, particularly in transportation and manufacturing of steel. As a result of these process improvements energy consumption in the production of

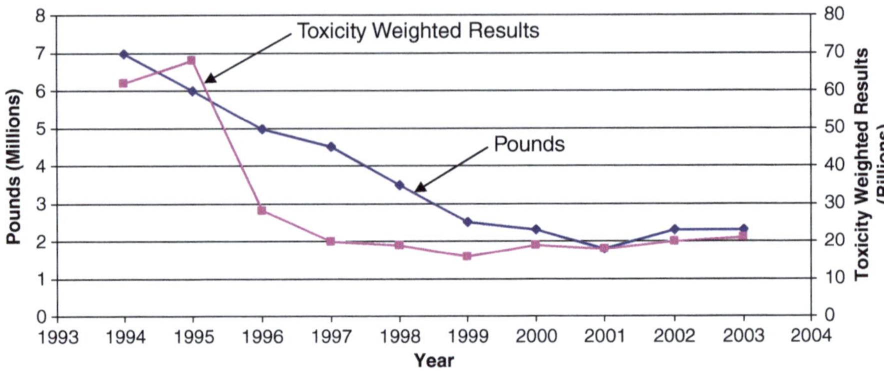

Fig. 2.3 Toxics Release Inventory (TRI) air and water releases by the Iron and Steel Sector (US EPA 2006)

Table 2.3 Total greenhouse gas emissions in teragrams of CO_2 equivalents (Calkins 2009)

Source	1990 (Tg CO_2 Eq.)	2005 (Tg CO_2 Eq.)	% Change
Carbon Dioxide (CO_2)			
Iron and steel production	84.9	45.2	–46.7
Aluminum production	6.8	4.2	–38.2
Zinc production	0.9	0.5	–44.4
Methane (CH_4)			
Iron and steel production	1.3	1	–23
Hydrofluorocarbons (HFCs), Perfluorocarbons (PFCs), & Sulfur Hexafluoride (SF_6)			
Aluminum production	18.5	3	–83.7

steel is continuing to decrease over time as shown in Fig. 2.2. These improvements were realized even before the sustainability movement because it made good business sense. With further funding in research of energy efficient technologies, the steel industry aspires to keep reducing their carbon footprint to become a carbon-neutral industry (Sustainable Steel: Energy Reduction 2010).

Use of timber in construction can have a significant impact on the environment, both positive and negative. Until third-party certified sustainably harvested wood became available, timber production resulted in soil erosion, pollutant runoff, increased CO_2 levels, and habitat loss. Timber that comes from certified forests has been managed to maximize timber yield, minimizing erosion to protect waterways, and promote healthy ecosystems for wildlife habitat. Unfortunately, only about 20% of the currently available timber products come from certified forests (DeStefano 2009). Also, to mitigate the effect of clear-cutting old growth forests, we rely on wood that is farmed using fertilizers and pesticides. These forests are harvested relatively frequently resulting in smaller diameter trees being used as structural wood, which can be produced into Engineering Wood (EW) products such as plywood, oriented-strand board panels, glued-laminated lumber, laminated and parallel strand lumber, and laminated veneer lumber. These EW products require binders, which will make EW products more difficult to recycle. The binder used in EW products is either phenol- or urea-formaldehyde resin. The Federal Green Construction Guide for Specifiers suggests avoiding products manufactured with urea-formaldehyde resin (a potential carcinogen). Phenol-formaldehyde is less harmful and is the primary binder used in the aforementioned EW products. The potential adverse effects of engineered lumber are offset by their more efficient use of natural resources and use of recycled content.

Also, a recent study by Kang and Kren (2007) indicates that structures account for 49% of sulfur dioxide emissions, 25% nitrous oxide emissions, and 10% of particulate emissions, all of which adversely affect urban air quality. A further impact that building construction has on the environment is solid waste from demolition, construction, and remodeling, which in 1996 was estimated at 136 million tons (approximately 35% of non-industrial waste) and 170 million tons by 2003 (US EPA 2009); this constitutes an increase of over 20% in 14 years. The average percentage breakdown of this solid waste by materials is as follows: 40–50% concrete and mixed rubble, 20–30% wood, 5–15% drywall, 1–10% asphalt roofing, 1–5% metals, 1–5% bricks, and 1–5% plastics. It is anticipated that in the near future, most of these materials will be reused or recycled.

2.4 Embodied Energy in Infrastructure Materials

More natural resources go into the built environment than any other manmade use. As resources become scarce (forests and fossil fuels), the infrastructure industry is resorting to recycling and reusing; this can also result in additional revenue and decreased costs in construction. Recycling is also promoted by state and local municipalities through increased fees in landfill use and waste emission reduction laws. The current state of the environment is slowly improving because of some of the sustainability practices being adopted by contractors, such as using locally produced and renewable materials and specifying materials with low embodied energy – energy required to produced, transport, and install the product.

At first glance structural engineers appear to have little to contribute to sustainability; however, Hays and Cocke (2009) make a compelling case to the contrary. One particular approach is to reuse existing buildings to reduce embodied energy in construction materials – after all, "the greenest building is one that is already built" Carl Elefante. Hays and Cocke (2009) presents an embodied energy analysis of an adaptive reuse of a 1950s, two-story concrete warehouse. To construct a new building and replace the 50,000 ft^2 space, it would require over 110 million BTU's worth of energy: 56.5 million BTU's embodied energy in the existing building, 0.775 million BTU's to demolish the building, and 56.5 million BTU's to construct the new building; a new high-efficiency building could not save this amount of energy over a 50-year service life (Hays and Cocke 2009). This example illustrates how the structural engineer can make a substantial contribution to sustainability.

Table 2.4 Benefits and challenges of deconstruction (Calkins 2009)

Benefits of deconstruction	Comments
Reduced environmental and health impacts from raw material use, acquisition, manufacture, and processing of new materials	Use of reclaimed or recycled materials will reduce virgin resource use, habitat destruction, energy use, and emissions from acquiring and manufacturing new materials.
Reduce landfill debris	Reduction in materials that must be landfill can save costs of landfill tipping fees, which are substantial in some areas. In some cases the savings will pay for the increased labor costs of deconstruction. Additionally reducing landfill disposal preserves land and may reduce the possibility of any future problems associated with landfill.
Management of hazardous resources	Deconstruction allows for management of hazardous materials, such as pressure treated-lumber, as they can be segregated and disposed of appropriately rather than commingled and land filled, where they hold the potential to leach hazardous substances. In addition, when hazardous waste is commingled with non-hazardous waste, the entire load must be treated as hazardous waste.
Strengthens the salvage and recycling industry	Growth in markets for reclaimed and recycled materials is directly related to the increase in deconstruction activities. The more deconstruction taking place, the stronger the markets for reused and recycled materials. This can result in job growth and benefits the local economy.
Design opportunities with use of reclaimed materials (e.g., aesthetic, historic, symbolic)	Reclaimed materials can add a layer of meaning to a project, revealing the cultural history of a place that is often difficult to achieve with mass-produced, internationally distributed, new materials. Reclaimed materials are sometimes unique and one of a kind.
Can achieve LEED credits	Deconstruction can contribute directly to achievement of two LEED credits: MR Credit 2.1: "Construction Waste Management: Divert 50% from Disposal" and "Divert 75% from Disposal." It can also help achieve several other credits in areas of "Material Reuse" and "Regional Materials."
Can save costs of new materials	Using reclaimed and reprocessed materials can often be cost effective, saving material acquisition expenses. Hauling and landfill expenses can be saved if materials are reused on site.

It may only be possible to reuse the entire building in a few cases; but every existing building has parts that can be reused or recycled. These options should be considered in every project in order to decrease the use of raw materials in the production of construction materials. There are a number of established programs that address reuse and recycle of materials: building deconstruction, construction materials salvage, and reuse of reclaimed materials in new site and building projects. Table 2.4 lists some benefits and opportunities associated with deconstruction (Calkins 2009). Use of recycled materials can decrease extraction of raw materials and reduce the embodied energy of a construction project, particularly for concrete and steel.

In order to reduce the embodied energy in concrete, the industry has instituted a number of programs. For example, concrete production methods have been improved, concrete is being obtained from local producers, and, in deconstruction cases, concrete can be recycled and used as aggregate. When concrete recycling and fly ash are used on the same project, the result is a decrease in the carbon footprint of concrete by as much as 50–60% (Teller and Bergman 2010).

As discussed in the last section, steel recycling and new processing methods have dramatically decreased its embodied energy (by as much as 75%) over the past six decades (see Fig. 2.2) resulting in steel having a relatively low impact on the environment. Due to its cradle-to-cradle (rather than cradle-to-grave) property, the majority of steel mills incorporate recycled steel into the production of new material; resulting in steel being, one of the most recycled materials today. In fact, at the end of a building's life, over 98% of all steel is recycled and reconstituted into new steel products (AISC 2011). Furthermore, technological advances in new machinery have allowed the steel industry to generate steel that is 40% stronger than steel from 40 years ago (Hewitt 2003). Along with the substantial reduction in greenhouse emissions discussed in the previous section, significant reduction in water utilization has been achieved, as much as 40%. The global steel industry continues to work to address sustainability issues by reducing steel's embodied energy and to further reduce its impact on the environment. In particular, they are working on higher strengths, better durability and sustainable characteristics to allow steel to become the most energy efficient construction material.

Much of the efforts from both the concrete and steel industries have been concentrated in reducing the effects of steel on the environment. Both industries continue to work on technologies that reduce their carbon footprint and embodied energy. These efforts should be rewarded through high marks on the LEED certification process for sustainable materials improvements. However, currently, LEED does not directly account for embodied energy, but it can be included indirectly in a number of the LEED areas. In future versions of LEED, embodied energy will probably be explicitly included; particularly through the work of the *Sustainable Preservation Coalition*, which has been formed to incorporate preservation, social, and cultural values into LEED (Hays and Cocke 2009).

2.5 Sustainable Attributes of Different Construction Materials

Wood and stone are considered to be the earliest construction materials and were used to build shelters by many ancient societies. Wood has continued to be used as a construction material since ancient times. Stone, on the other hand, was replaced with mud brick in some parts of the world. Mud brick can be considered the earliest composite material because it was reinforced with natural straw to prevent cracking of the brick, given the poor tensile properties of dried soil. This practice was common at the time of ancient Egypt, and was so successful that it is still in use in many parts of the world today. The early practice of reinforcing mud with straw is

Table 2.5 Net Carbon emissions in producing a ton of various materials (Falk 2010)

Material	Net CO_2 emissions (ton CO_2/ton)[a,b]	Near-term Net CO_2 emissions including CO_2 storage within material (ton CO_2/ton)[c,d]
Framing lumber	0.033	–0.457
Medium-density fiberboard (virgin fiber)	0.060	–0.382
Brick	0.088	0.088
Glass	0.154	0.154
Recycled steel (100% from scrap)	0.220	0.220
Concrete	0.265	0.265
Concrete[e]	0.291	0.291
Recycled aluminum (100% recycled content)	0.309	0.309
Steel (virgin)	0.694	0.694
Plastics	2.502	2.502
Aluminum (virgin)	4.532	4.532

[a]Values are based on life-cycle assessment and included gathering and processing of raw materials, primary and secondary processing, and transportation

[b]Source: EPA (2006)

[c]From Bowyer and others (2008); a carbon content of 49% is assumed for wood

[d]The carbon stored within wood will eventually be emitted back to the atmosphere and the end of the useful life of the wood product

[e]Derived based on EPA value for concrete and consideration of additional steps involved in making blocks

documented in the Bible, Pharaoh told the Hebrew slaves that they would not be given any straw, yet they would have to produce their full quota of bricks. The ancient Romans invented concrete and built many structures and monuments with concrete, including the Pantheon in Rome, which was constructed using concrete over 2,000 years ago and stands to this day. Steel is a more recent discovery, with the first application to infrastructure happening in 1779, the Coalbrookdale Arch Bridge in Shropshire, an iron bridge on England's Severn River. This bridge weighs 378 ton and spans 100 ft, and is still used as a pedestrian bridge today. In the past century, a number of new construction materials were invented, from high-performance conventional materials to advanced composite materials. As civilization has progressed, the embodied energy (as well as the CO_2 emissions) in construction materials has continued to grow over the centuries. As shown in Table 2.5, the CO_2 emissions per ton of material has also grown as new materials have been developed, with virgin aluminum being the worst at 4.5 ton of CO_2 per ton of aluminum. Following is a discussion of the steps currently being undertaken to reduce the embodied energy and to promote sustainability in contemporary construction materials: concrete, steel, timber, masonry, and composites.

2.5.1 Concrete

One of the key ingredients in the production of concrete is cement (the other two are aggregate and water). Cement production is energy intensive and generates large amounts of CO_2 – as noted earlier, producing a ton of cement generates approximately 1 ton of CO_2. As cement producers are under increased pressure to reduce CO_2 emissions, they have resorted to optimizing cement mixtures and adding complementary cementing materials, such as recycled fly ash (a pozzolanic by-product of coal-fired electricity generation) and blast-furnace slag (a pozzolanic by-product of steel blast furnaces). The benefits of using these pozzolanic by-products are two fold: the materials are diverted from the landfills and the cement's environmental impact is reduced because of its replacement with a carbon-neutral by-product. According to the Portland Cement Association, fly ash and slag are optimal substitutes for raw cement because their use results in no degradation to the mechanical properties of concrete. This is particularly important since cement on average accounts for 83% of the energy needed to produce concrete, but only makes up approximately 15% of the mix. Therefore, using these by-products and reusing some hardened concrete can result in a potential for creating a material that is 50–60% recycled and much more energy efficient (Teller and Bergman 2010).

Paul Tennis from the Portland Cement Association (PCA) has noted that as much as 20% of cement never hydrates because rapid reaction inhibits uniform cement distribution throughout the concrete material (Arbuckle 2010). This issue is more critical for early strength concrete because concrete may not reach its full potential compressive strength at the specified age. In efforts to minimize the un-hydrated cement, PCA has developed a number of admixtures to promote cement hydration and obtain a higher compressive strength earlier, resulting in a high performance material. High performance concrete also requires selection of high quality aggregate (sand and gravel) with optimized gradations to produce a more impermeable hardened concrete. Because of lower concrete permeability, corrosive chemicals (such as chloride) are less likely to reach the concrete reinforcement, postponing corrosion of the steel, which results in an increased material longevity (durability). Thus, mix designs optimized for early strength and rapid construction can lessen concrete's environmental impact. Improving concrete's durability can also result in a much lower life-cycle cost. An added beneficial sustainable characteristic resulting from higher strength concrete is smaller components for a given target capacity.

From the discussion in this section, it appears that concrete's adverse environmental impacts can be largely attributable to cement production. As discussed earlier, and noted in Table 2.2, PCA allows replacing as much as 40% of cement with fly ash in a given concrete mix, making concrete relatively competitive to other construction materials from a sustainability stand point. Until recently, the main impediment to reducing cement content in concrete was conventional industry standards, which had institutionalized cement intensive mix designs that exhibited poor long-term performance resulting in unnecessary adverse environmental impacts. However, since the concrete industry is organized around consensus standards and most professionals now recognize the importance of sustainability, sustainable practices are slowly being incorporated into the concrete industry. In fact, fly ash and slag ASTM standards (ASTM C618 and ASTM C989, respectively) have been developed as a testament to the concrete industry's commitment to sustainability. The steps being undertaken by the cement industry to reduce its carbon-footprint, particularly the use of complementary cementing materials, will result in concretes with lower embodied energy, lower carbon emissions, lower environmental impact from the extraction and processing of virgin materials, and an increase diversion of by-product materials from landfills; all of which will result in a more durable concrete material with a much longer service life and lower overall cost.

Concrete can also lessen its environmental impact by incorporating substitute recycled materials for natural coarse or fine aggregates, including: crushed blast furnace slag, brick, glass, foundry sand, granulated plastics, waste fiberglass, sintered sludge, and mineralized wood shavings (Calkins 2009). This would result in a reduction of virgin aggregates in concrete and should divert these waste materials from landfills.

2.5.2 Steel

Steel is typically produced by one of two methods: electric-arc furnace (EAF) or basic oxygen furnace (BOF). Each uses different amounts of recycled material; in the United States, steel produced by the EAF process typically contains 90% or more recycled content, while steel produced by the BOF process contains only approximately 25% recycled content (Hewitt 2003). The two processes also use different sources of energy: EAF uses electricity while BOF uses coal or natural gas. Therefore, we can conclude that EAF is a more environmentally friendly process than BOF. Fortunately, most structural steel (all of the wide flange shapes) are produced using EAF; while steel produced using the BOF is predominantly used in non-structural applications, except for Hollow Structural Shapes (HSS) and steel decking. So designers seeking a sustainable building should try to minimize the use of HSS members. The remainder of this section emphasizes structural steel that is produced by the EAF method.

The structural steel industry as a whole has made tremendous progress towards some of the most important goals of sustainability; particularly promoting energy efficiency, reducing the use of virgin materials, minimizing site disturbance, and providing a healthier living environment. Steel is 100% recyclable, in fact, it can be multi-cycled without any loss of quality – that is, steel can be recycled multiple times, making it truly a cradle-to-cradle material. Currently, steel produced in the United States contains on average 93.3% recycled steel scrap (which is the highest recycled content of any building framing material), and every time steel is recycled there is no degradation to its mechanical properties, unlike other materials that can only recycled into a lower quality product (down-cycled) (AISC 2011). Because of the increase in steel recycling and changes to the production process, the raw materials and energy (including CO_2 emissions) previously used in the production of steel have drastically decreased over the past 30 years. And given that most steel is produced using electricity, the embodied energy in steel will continue to decrease as electricity continues to come from more renewable sources, such as solar and wind; eventually allowing the steel industry to attain its goal of producing steel that has no carbon footprint (zero-carbon steel).

Like with concrete, the use of recycled steel increases the volume of materials diverted from the landfills and repurposed; by-product materials that would typically result from extracting raw material from the ground are also diverted from the landfills since virgin steel is replaced by recycled steel. In fact, minimal steel waste is generated at fabrication facilities or construction sites; most waste generated is recycled. Also, when considering that steel buildings require minimal ongoing maintenance, have long lives, and are recyclable at the end of their life, the resulting environmental impact of steel is minimal. Furthermore, there is a potential for steel material reuse, particularly steel from deconstruction, which is made easier and faster because of use of steel bolts to fasten steel systems together.

A number of advances in recent years have contributed to the reduction in the volume of steel used in a given project. In particular, high strength steel, which has 40% higher strength (36–50 ksi) than steel used 30 years ago (Hewitt 2003). This increase in strength results in smaller elements, which in turn result in smaller supporting superstructure and foundation components.

Since steel has a high strength-to-weight ratio, it can carry large loads with a smaller footprint of the structural system in multi-story buildings. This reduces the impact of the building on the site by requiring less widespread site development. A smaller building footprint will also have a lower impact on the site by increasing the remaining size of the site and making it available for permeable ground cover to provide for storm water infiltration.

Other innovations in the design of steel buildings that can have a significant impact on their sustainability include the use of a design-build approach, which entails reducing the delivery schedule of a project by overlapping the design and construction phases. This compressed schedule also lessens the adverse effects of construction on the site. A new innovation in building design known as Building Information Modeling (BIM) allows the designer to create 3-D computer models of all buildings systems to determine their interactions and conduct conflict resolutions before construction even begins. BIM can be used to optimize the integration of mechanical systems within the floor beams leading to lower floor-to-floor heights, which results in less building volume to be heated or cooled (lowering the building's energy consumption).

Continuous improvement in water resource management in the production of steel has resulted in a 95% water recycling rate; currently less than 70 gal of water are consumed per ton of steel produced (AISC 2011). In the areas of indoor environmental quality, steel framing systems can be used to span large indoor areas to improve occupant comfort, and span large wall openings for windows to allow natural lighting, which can result in a reduction in electrical consumption and further reduction in CO_2 emissions.

The steps already taken and steps planned by the steel industry to reduce its carbon-footprint, particularly using high-strength steel, new innovative design techniques, and recycling and reusing steel, will result in buildings with a lower embodied energy, lower carbon emissions, lower environmental impact from the extraction and processing of virgin materials, and an increase diversion of by-product materials from landfills; all of which will result in a more sustainable and cost effective material.

2.5.3 *Timber*

Timber is a renewable resource, particularly timber from wood that comes from certified forests. Also, because of reforestation practices, the forested area in North America is the same size as 100 years ago (Ward 2010). The net annual growth is 3% larger than the harvests and other losses combined. Furthermore, on average 98% of a given tree brought to a mill is used as timber, paper, Engineering Wood products, or as fuel for bio-energy (Ward 2010). When considering these factors and the fact that timber actually sequesters CO_2, this material can be considered a carbon negative material at least in the short-term (see Table 2.5). That is, wood that comes from managed forests removes carbon from the atmosphere through photosynthesis (Falk 2010):

$$\text{Energy(sunlight)} + 6H_2O + 6CO_2 \rightarrow C_6H_{12}O_6 + 6O_2$$

This makes timber the most effective technology for extracting CO_2 from the atmosphere. Not only is a substantial amount of CO_2 sequestered, but oxygen (O_2) is released; for every 1.47 lb of CO_2 removed from the atmosphere, 1.07 lb of O_2 are released (DeStefano 2009). Of course, if the timber is burned, or decomposes the process is reversed, releasing the CO_2, yielding a net carbon emissions, which is still the lowest of all construction materials, as shown in Table 2.5.

Because wood requires low energy during processing and since more than 60% of this energy comes from bio-fuel (a carbon-neutral energy source), the embodied energy in timber is much lower than other construction materials (Falk 2010). Half of the energy required to produce lumber goes into drying the wood in a kiln, so specifying green (un-dried) lumber is 50% more energy efficient. EW products also have a relatively low embodied energy; though higher than sawn lumber – more processing involved in EW. However, EW products, such as oriented strand board and composite lumber, use wood chips manufactured using smaller trees from shorter rotation harvests, which sequester more CO_2 than those in longer-rotation forests (Falk 2010). This balances some of the additional embodied energy required to process EW products.

One of the main benefits of wood from an environmental stand point is that it is biodegradable. Unfortunately, in exposed and high moisture content applications the wood is vulnerable to decay or attack from wood destroying insects, which must be repelled using preservatives to ensure a long-term durability. Recently a number of non-toxic preservatives, mechanical barriers and coatings have been developed to address wood durability. Other durable material alternatives to treated sawn lumber include components fabricated from EW products. Also, an alternative to preservatives is the use of decay-resistant wood species, such as Redwood, Cedar, and Cypress; particularly if these species are locally available.

Local availability of wood species is a major advantage of timber over other construction materials. A designer should always specify locally or regionally available species: both for sustainable and economic reasons. The following are the most widely available species in different parts of the United States: Douglas Fir in the West, Southern Yellow Pine in the Southeast, and Spruce-Pine-Fir in the Northeast and Midwest (DeStefano 2009).

Unlike steel, where recycling of pre- and post-consumer steel for structural applications is common place, timber recycling is primarily used as bio-fuel. However, timber reuse from deconstruction is possible, though more complicated than for steel. Reusing post-consumer recycled timber for structural purposes requires the timber to be re-graded according to standardized grading rules, and in some cases it may have to be tested. Also, recycled wood can be processed into landscape mulch, which is useful in retaining moisture in the soil, lessening the water demands from plants in the landscape.

There are a number of innovative developments in the timber industry, which can result in more efficient and less wasteful practices. One particular development is the use of advanced framing techniques to optimize the layout for efficient materials use, particularly in light framed wood construction. For example, one design strategy is to layout the building so that dimensions are multiples of 2 ft, which can accommodate plywood sheets' typical sizes and allow placement of joists and studs at 24 in on center. Using 24-in spacing instead of the common 16-in spacing should reduce the number of studs and joists needed. This will also allow the alignment of horizontal framing members (joists) with studs, which can also eliminate the need for double member top-plates. Though this strategy may result in fewer pieces, it may require an increase in the size of individual components; increasing the stud sizes from 2 × 4 to 2 × 6 and wall sheeting from ½- to ⅝-in (DeStefano 2009). Also, material strength should be used to its full approved capacity, or the most efficient grade of lumber to suit the structural demand should be specified. Other strategies include designing headers (including those for windows and non-structural walls) to the required structural capacity, and using double 2× lumbers instead of 4× wherever possible. These strategies should be carefully considered in light of the project structural demands, durability objectives, and design fee.

The timber industry appears to have made the most progress towards becoming carbon neutral thus far. This industry is expected to continue to reduce their carbon footprint by promoting sustainably managed forest, use of non-toxic preservatives, use of locally available wood species, development of durable materials and components in the form of EW products, use of advanced framing techniques, design of structures that can be altered or adapted to new uses or loading conditions, and by requiring that construction site waste and demolition debris be sorted and recycled or used as bio-fuel.

2.5.4 Masonry

For the purpose of this discussion, we will divide masonry into categories based on application (structural or non-structural) and material (concrete, clay, and fly ash). In structural applications, masonry is used primarily as walls, which can serve as combined gravity and lateral load-bearing elements. In order to serve this purpose, masonry walls must be reinforced with steel – though many un-reinforced clay masonry walls were built in the past and are still standing. Masonry in the form of concrete masonry units (CMU) is the most common form of masonry used in structural walls, while brick (clay, concrete, or fly ash) is almost exclusively used for facing buildings. Prefabricated CMU cellular elements serve as formwork for the concrete walls, eliminating the need to use timber formwork. This however is increasingly uncommon because the process of assembling a masonry wall is labor intensive and more efficient methods have been developed to construct concrete walls (including site cast, tilt-up, or precast walls).

As for non-structural applications, exterior facing masonry walls can serve to provide thermal mass, which, depending on the thickness of the wall, can markedly improve the thermal performance of a building (Volz and Stovner 2010a). The process entails absorbing energy from the sun during the daylight hours and releasing it as radiant heat at night; a process which lessens the effect of temperature swings within the building envelope. And for much of the year (in many parts of the world) maintains thermal comfort without the need for heating or air-conditioning. Masonry is also used to improve the acoustic comfort of buildings, where material mass is the primary means to decrease sound transmission. When masonry is used as thermal or acoustic mass and as structural walls to produce a sustainable synergy, more material may be needed than is required for strength.

From a sustainability standpoint, masonry can also be used in permeable pavement applications. Masonry (concrete, clay, or fly ash) units can improve the storm water management of a site by providing a permeable material for storm water infiltration over a hard surface. This type of permeable pavement system can also reduce surface runoff by improving the soil percolation of a site.

The embedded energy in masonry is considerably different for concrete and other brick materials. Like regular concrete, CMU and brick have an embedded energy and a carbon footprint that can be minimized by substituting cement with complementary cementing materials, such as fly ash. Clay brick is a fired material that requires large amounts of energy in its production, which results in a large embodied energy and carbon footprint. To address this last issue, the masonry industry has developed fly ash brick, which has comparable or better properties than fired clay brick (Volz and Stovner 2010b). A comparison of the embodied energy for all masonry materials is given in Table 2.6; notice that fly ash brick has approximately half of the embodied energy of concrete materials and only 15% of fired clay brick. Also, fly ash has better insulation properties compared to fired clay brick (Volz and Stovner 2010a).

Table 2.6 Embodied energy in masonry materials (Volz and Stovner 2010a)

Material	Embodied energy ($MBtu/yd^3$)
Concrete masonry units	0.962
Concrete brick	0.946
Fired clay brick	3.28[a]
Fly ash brick	0.492[b]

[a]This value is given by the Brick Industry Association, NIST estimates it at 6.71 $MBtu/yd^3$
[b]This value is estimated as 15% of fired clay brick (Volz and Stovner 2010b)

Masonry has been very popular as a construction material for generations and remains one of the most popular facing materials for buildings. Many vintage buildings are being retrofitted for reuse as refurbished space for residential or commercial purposes. In some cases, advance composites are used to bring these buildings up to code and to extend their useful life well pass their intended life. As Carl Elefante put it, "the greenest building is one that is already built". Recycled masonry can also be crushed and used as fill material or to fabricate new brick.

The masonry industry has made great progress in reducing the embodied energy of its products; particularly that of brick. Brick continues to be one of the most popular building facing materials because of its classic beauty, high thermal and acoustic mass, and durability. The industry is expected to continue promoting practices that will reduce its carbon-footprint; such practices include: minimizing the use of cement in CMU, replacing fired clay brick with more sustainable fly ash brick, and reducing masonry waste by reclaiming block and brick materials from the waste stream to be crushed for use as fill material.

2.5.5 *Advanced Composite Materials*

Advanced composite materials have had very limited used in infrastructure applications. In fact, they are typically only used as retrofit components or components targeted to a specific issue, such as corrosion in steel concrete reinforcement. The main advantages of composites over their conventional counterparts are high structural performance, high specific mechanical properties, and durability. The most common applications for composites to date include: rehabilitation of structures, seismic retrofitting of columns, and bridge decks (both whole and FRP reinforced concrete). All these applications of composites are intended to extend the life of structural systems well beyond their expected life, which may, in many cases, balance their adverse environmental effects. As can be seen from Table 2.5, plastics (a key ingredient of composites and typically well over half, in many cases as much as 75% – BRE and NetComposites 2004) have a very large carbon footprint, second only to aluminum derived from virgin material. However, a recent study based on life-cycle analysis of structural components fabricated by Strongwell Composites indicates that the embodied energy of some composite components is lower than that of steel members made from virgin materials (Black 2010). The report speculates that this is primarily due to the composites' superior specific properties, such as high strength-to-weight and high stiffness-to-weight ratios. However, when compared to aluminum or steel parts made from average recycled content, composite parts' embodied energy is comparable to that of metal materials. Timber on the other hand, the report concludes has a marked advantage over composites, approximately 50% lower embodied energy footprint compared to composites. One way to lessen the environmental impact of composites is by increasing fiber content, which would decrease the percentage of resin required (Black 2010); unfortunately, this would also increase the cost.

One other area where composites stand to have a significant positive impact on the environment is renewable energy systems, particularly wind power. To harness the power from wind, large turbines are placed on high towers. The turbine blades must be relatively light for transportation and efficient operation, which is why most of the turbine blades are manufactured using composites. As of 2007, more than 17,000 turbines (nearly 50,000 blades) had been installed around the world, for a total capacity of 94,112 MW; this constitutes the largest single applications of engineered composites in the world (Hollaway 2010). A detailed discussion of other applications of composites in civil infrastructure can be found in the following chapters and in Hollaway (2010).

2.6 Building Codes, Specifications, and Standards

Building codes and industry standards are currently undergoing changes to reflect sustainability principles and practices in order to lessen civil infrastructure's adverse impacts on the environment and to address recent changes in the law; e.g., California's Title 24 – *Energy Efficiency Standards for Residential and Non-residential Buildings*, which will require net-zero energy construction of homes by 2020 and 2030 for commercial buildings. This standard is also known as "CALGREEN" and became effective on January 1, 2011. Even before CALGREEN, California required all new construction (or renovations) of state-owned buildings to meet or exceed LEED Silver level standards. Other states have enacted similar laws for public buildings; in fact, over 30 states now require that public buildings meet or exceed sustainability standards comparable to LEED certification level (as of early 2011). Also, over 200 local governments in over 40 different states have adopted LEED certification mandates for public buildings (www.USGBC.org). A number of federal agencies and departments have also adopted sustainability standards comparable to LEED certification.

Professional organizations are developing policy statements to achieve the same goal; e.g., the *American Institute of Architects (AIA) 2030 Commitment* calls for all buildings to be carbon-neutral by 2030. The American Society of Civil Engineers began addressing sustainability as far back as 1996, when it revised the Code of Ethics to address sustainability and embrace it as an ethical obligation; ASCE also developed *Policy Statement 418 – the Role of the Civil Engineer in Sustainable Development* to reiterate the leadership role civil engineers must play in sustainable development. Also, ASCE has established a *Sustainability Committee*, which was formed to educate civil engineers on how to promote sustainability practices in civil infrastructure design. This committee has developed a *Sustainability Action Plan* in order to "define the role of society and the profession of civil engineering in advancing sustainability in civil infrastructure" (ASCE 2010). This group has also collaborated with other organizations including the U. S. Green Building Council and Engineers without Borders to educate engineers on sustainable construction practices.

ASCE, along with the American Council of Engineering Companies (ACEC) and the American Public Works Association (APWA), has recently established the *Institute for Sustainable Infrastructure*, which has a mission "to provide sustainability products and services that will transform infrastructure design, construction, and operation by taking into account the 'Triple Bottom Line' – economic, environmental and social impacts." The institute's first priority was to develop a new rating system focused on civil infrastructure, the Sustainability Ratings for Engineering Projects (http://www.asce.org/Sustainability/ISI-Rating-System/). "The rating system will be scalable to accommodate projects of all sizes and complexities, adaptable for specific needs and circumstances, and offer performance-based outcomes rather than being prescriptive." Given that no other U.S. programs currently provide a system to evaluate the sustainability of many infrastructure systems, this comprehensive rating system constitutes a major step forward towards creating a truly sustainable civil infrastructure.

The Structural Engineering Institute (SEI) of ASCE has also recently established the *SEI Sustainability Committee* to advance the "understanding of sustainability in the structural community and incorporate concepts of sustainability into structural engineering standards and practices." This group also recently published *Sustainability Guidelines for Structural Engineers* (Kestner et al. 2010) to highlight the significant environmental impacts of structural materials. As an advocate for sustainable structures, *SEI Sustainability Committee* recently developed public comments to address the decision by the U. S. Green Building Council to eliminate structural materials from the 2012 draft LEED rating system, particularly credits for Regional Materials, Recycled Content Materials, and Bio-Based Materials. It is anticipated that in the final LEED credits structural materials will be included because of the work of this group.

A number of new sustainability standards have recently been published. The American Society of Heating, Refrigerating, and Air-Conditioning Engineers (ASHRAE), in conjunction with the U.S. Green Building Council and the Illuminating Engineering Society of North America, has developed *ASHRAE 189.1, Standard for the Design of High Performance, Green Buildings except Low-Rise Residential Buildings.* ASTM International has formed a committee, *E60 – Committee on Sustainability*, to develop sustainability-related standards to promote "consistency in sustainability across mainstream markets." This committee is divided into subcommittees, one of which is *Committee E60.01 on Buildings and Construction*. This subcommittee is responsible for maintaining a number of standards, including *ASTM E1991 – 05 Standard Guide for Environmental Life Cycle Assessment (LCA) of Building Materials/Products*, which is used to quantify environmental impact, a key metrics of sustainability.

Materials specifications and standards such as those maintained by the leading materials organizations (ACI and PCA for concrete, AISC for steel, and AF&PA for timber) typically addresses life safety values; not public health and welfare, both of which are included in building codes, in addition to life safety. However, all these organizations are currently developing strategies to address and incorporate sustainability in structural design practices. As a first step, these organizations (ACI/PCA, AISC, and AF&PA) have formed working committees charged with addressing sustainability in their respective specifications and standards.

The American Concrete Institute for example has established *ACI Committee 130 – Sustainability of Concrete*, which has a mission to "develop and report information on the sustainability of concrete." This committee has established a subcommittee to address sustainability in codes and standards, *ACI Committee 130-0E – Design/Specifications/Codes/Regulations*. In 2004, ACI Committee 130 developed a cooperative partnership with the U. S. Green Building Institute (USGBC) to improve the perception of concrete relative to sustainability (Arbuckle 2010). Furthermore, the combined efforts of ACI and the Portland Cement Association have resulted in several standards that have greatly reduced the impact of the cement industry on the environment; in particular, the development of ASTM C618-08a for fly ash and ASTM C989-09a for slag standards. Another important standard developed by ACI is the *Guide for the Design and Construction of Externally Bonded FRP Systems for Strengthening Concrete Structures* (ACI 2002), which is a common process for retrofitting structures using composite materials.

The members of the International Steelwork Contractors Group (ISCG), Australia (ASI), Canada (CISC), New Zealand (SCNZ), South Africa (SAISC), United Kingdom (BCSA), and the U. S. the American Institute of Steel Construction (AISC), have established the *Steel Sustainability Task Group* (SSTG) to address sustainability issues related to steel construction. At their first meeting in 2007, the group established the following set of objectives (www.AISC.org):

- To use the international framework provided by the ISCG to develop steel as a sustainable form of construction in terms of economic viability, social progress, and environmental responsibility.
- To collaborate on defining, developing, and measuring the sustainable credentials of structural steel.
- To monitor competitive trends and claims concerning the sustainable credentials of other structural materials, and to share this information with other ISCG members.

The group's focus is on minimizing the carbon footprint of the steel industry, particularly for manufacture of virgin steel from iron ore. Also, AISC along with the

American Iron and Steel Institute (AISI) are involved in the development of voluntary and model codes and standards; AISC has a voting member representative on the ASHRAE committee and AISI is represented on the IGCC working group (www.AISC.org).

The American Forest & Paper Association (AF&PA), which maintains the National Design Standards for timber, has embraced sustainability and made it part of the entire organization. AF&PA specifies the following practices as those that define sustainability for the forest products industry:

- Sustainable use of renewable resources.
- Leading the way in recycling.
- Reducing our environmental footprint.
- Generating and conserving energy and materials.
- Reducing greenhouse gases.
- Benefiting the communities, employees, and families everywhere our products are made and sold.

AF&PA has worked with the Forest Stewardship Council (FSC) (the only one currently recognized by the USGBC-LEED certification program) and other third-party non-profit organizations that certify wood from sustainably managed forests. This certification program provides timber designers the opportunity to specify timber that meets the standards of the green building community.

Also, organizations in charge of developing model building codes, such as the International Codes Council (ICC) – developer of the model code *International Building Code* (IBC) – have been working with third-party consensus voluntary standards organizations such as the USGBC to develop model "green" codes for residential and commercial buildings. ICC and USGBC recently published a draft of the first version of a "green" building model code, the *International Green Construction Code* (IGCC). ICC plans to publish the code in 2012, which is expected to be consistent and harmonized with the family of codes maintained by ICC. IGCC substantially incorporates the ASHRAE 189.1 provisions. A model "green" code for residential construction was recently published by the National Association of Home Builders, the *National Green Building Standard* (NGBS), see www.nahbgreen.org. A comparison of these "green" model building codes (IGCC and NGBS) to the USGBC-LEED voluntary standards is given in Table 2.6. Notice that all three documents (as well as ASHRAE) contain sections on materials, which are maintained by committees that include representatives from all the organizations representing the different categories of building materials (concrete, steel, and timber). Further input on the development of these codes and standards is obtained from all interested parties through a consensus process of evaluating proposed guidelines. As can be seeing from Fig. 2.4, with the advent of model "green" building codes and standards (Table 2.7) progress towards a neutral environmental impact infrastructure is achievable in the foreseeable future.

This progress has been made possible by the innovation of voluntary guidelines, such as LEED, which will soon be enforced, or at least enforceable, because of "green" codes. These voluntary guidelines have allowed the infrastructure industry

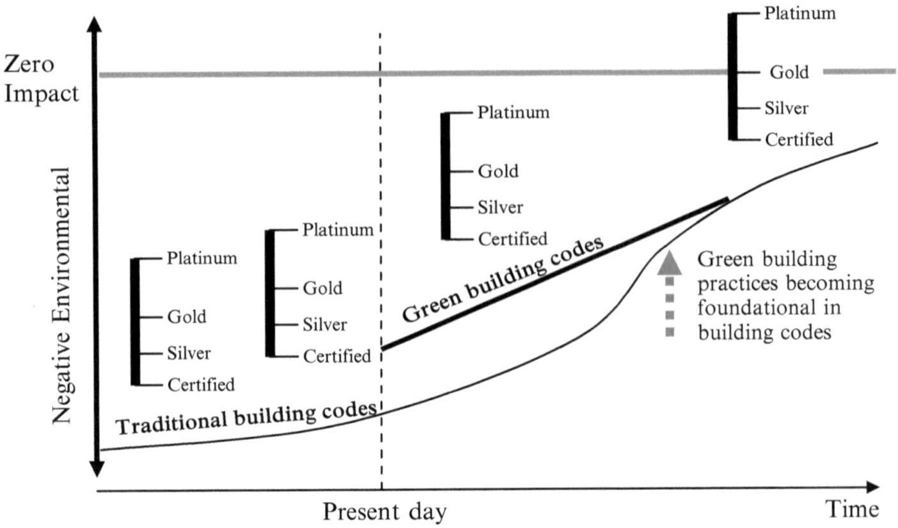

Fig. 2.4 Progress toward sustainability (USGBC 2010)

Table 2.7 Comparison of green building codes to LEED

LEED v3 (2009)	ICC-IGCC	NAHB-NGBS
Sustainable sites	Chapter 4. Site development and land use Chapter 11. Existing buildings site development	Chapter 4. Site design and development Chapter 5. Lot design, preparation, and development
Water efficiency	Chapter 7. Water resource conservation and efficiency	Chapter 8. Water efficiency
Energy & atmosphere	Chapter 6. Energy conservation, efficiency and atmospheric quality	Chapter 7. Energy efficiency
Materials & resources	Chapter 5. Material resource conservation and efficiency Chapter 10. Existing buildings	Chapter 6. Resource efficiency
Indoor environmental quality	Chapter 8. Indoor environmental quality and comfort	Chapter 9. Indoor environmental quality
Locations & linkages		
Awareness & education	Chapter 9. Commissioning, operation and maintenance	Chapter 10. Operation, maintenance, and building owner education
Innovation in design		
Regional priority		

to experiment with new sustainable methods, procedures, and products that have already begun transforming many aspects of infrastructure design, construction, and operation. In fact, LEED was designed to encourage and accelerate development and adoption of sustainable practices in building construction by creating and implementing a universally understood and accepted standard to verify buildings as sustainable.

All though LEED is currently the most common approach to quantify the level of sustainability a building can achieve, other groups have developed methods to measure sustainability of buildings. The Green Building Initiative (GBI) has developed and administers Green Globes, which is an online assessment protocol and a rating system for green building design, operation, and management. Green Globes ranks buildings on a scale of 1–4 globes (www.thegbi.org). The Green Building Initiative recently completed a new American National Standard (ANSI) for commercial "green" buildings, the first of its kind – *ANSI/GBI 01–2010: Green Building Assessment Protocol for Commercial Buildings.* The standard is intended to provide additional tools to the building construction industry by providing a user-friendly list of credible third-party certification programs for products and materials that qualify as sustainable.

2.7 Bridges

There are currently no national standards to measure or rank the level of sustainability of a bridge, similar to the USGBC LEED benchmark for buildings. However, it would not be difficult to develop a ranking system comparable to the LEED standard to quantify bridge sustainability. In fact, only the LEED Indoor Environmental Quality category could not be matched to a sustainable bridge metric. The others (in particular, Materials and Resources as well as Innovation in Design) can easily be translated to some equivalent sustainable bridge design goals to provide guidelines to designers who want to quantify the sustainable performance of their new, or retrofit, bridge projects. A first step in the development of guidelines for designers is given by Whittemore (2010); in particular, he provides a list of questions to guide designers in four areas: Sustainable Sites, Water Use and Quality, Energy and Transportation, and Materials and Resources. One area that is not mentioned in the article is Innovation in Design, which can result in tremendous advances in sustainability.

The Director of the Federal Highway Administration (FHWA) Office of Bridge Technology, Myint Lwin has publicly supported the concept of "Green Bridges" and the development of green standards for bridge construction (Lwin 2008). Lwin states that "In 'Green Bridges,' the design, construction, and maintenance practices should give full consideration to at least the following areas:

- Attention to safety, durability, mobility (traffic flow), and efficiency;
- Compliance with environmental and preservation laws and regulations;

- Application of context sensitive solutions;
- Sustainable site selection and planning;
- Utilization of high performance materials and quality workmanship;
- Conservation of materials and resources; and
- Avoidance of negative impacts on the ecosystems."

Considering the large mounts of materials that go into bridges, "Green Bridges" should have a rather large, positive impact on the environment. Also, a number of the chapters in this book will discuss innovative technologies that can be used in the design and retrofit of bridges; in particular, the use of composite materials to extend the service life of bridges.

2.8 Performance-Based Design (PBD) and Its Role in Sustainability

Current building codes and standards only address the life-safety function of buildings in the event of rare natural disasters, which represents a significant shortcoming from a sustainability standpoint. Sustainability in its current definition is more closely related to serviceability (that is, buildings should perform all original intended functions) than life-safety. Allowing a building to be serviceable after going through a rare natural disaster decreases the total building cost and environmental impact over its lifetime. Therefore, sustainability may be embedded in design codes and standards by decreasing the life-cycle cost of a building. This can be accomplished through Performance-Based Design (PBD), which requires that buildings perform to an acceptable (specified) level of serviceability and safety for different levels/risks of extreme events (such as earthquakes). PBD departs from previous design philosophies (Allowable Stress Design [ASD] and Load and Resistance Factor Design [LRFD]) in a fundamental way; PBD accounts for the consequences associated with hazards (demand) and vulnerabilities (capacity), where consequences can be measured in different ways including monetary cost and casualties (Tang et al. 2008). This new design philosophy requires designers to go beyond code specifications and accurately predict how structures will respond during extreme events. For seismic-based design for example, the following target-based performances should be considered:

1. Fully Operational – Facility continues in operation with no significant damage to structural or non-structural components.
2. Operational – Facility continues in operation with minor damage and minor disruption in nonessential services.
3. Life Safe – Life safety is substantially protected, damage is moderate to extensive.
4. Near Collapse – Life safety is at risk and damage is severe, but structural collapse is prevented.

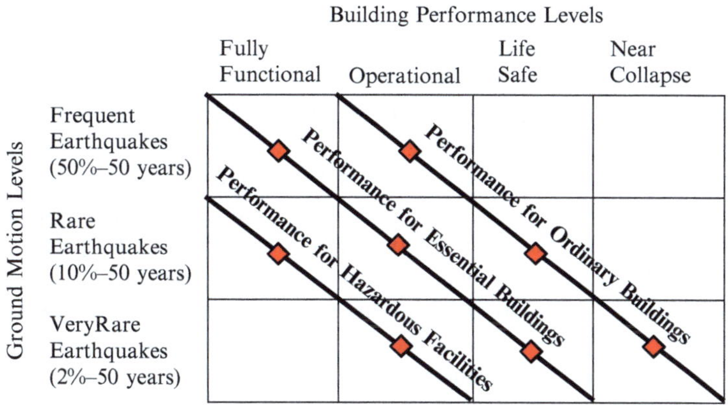

Fig. 2.5 VISION 2000 performance objectives (Hamburger 1997)

Buildings with different levels of importance and functionality will be assigned different performance requirements. Figure 2.5 depicts the performance requirements of various building types.

In a typical building, the structural system usually accounts for 10–20% of the construction costs (Kneer and Maclise 2008). Since PBD requires more complex analysis than conventional design philosophies, the cost of the structural system is as much as 10% higher for buildings constructed on a performance basis. This only translates into a 1–2% increase in the overall cost of the building. Therefore, an initial investment in a performance-based designed structural system can minimize or eliminate the potential massive costs resulting from an un-operational building after a natural disaster while repairs are being performed.

As previously discussed, sustainability first began as a term for being environmentally cognizant, but over the years has transformed into a methodology to evaluate how well a building can be maintained, how to minimize a building's impact on the environment, and how to minimize a building's overall life cycle cost. Given that the costs associated with the high-performance architectural, mechanical, electrical, and plumbing systems are significantly higher than for conventional buildings, protecting them with a high performance structural systems would result in significantly lower life-cycle cost. Therefore, protecting these high-performance systems is essential; particularly for buildings likely to experience the effects of a natural hazard over their lifetimes. Kneer and Maclise (2008) noted that most damage after an earthquake is to the non-structural components of a building. As stated earlier, in the latest version of LEED a series of bonus points have been introduced in Innovation & Design. By considering the material cost associated with repairing high-performance systems and replacing building contents for various levels of hazards, PBD can be utilized to show a quantifiable material waste reduction in the event of a natural hazard, such as an earthquake.

Current trends indicate that PBD will become widely accepted from a sustainability standpoint because it reduces material replacement and lowers life-cycle cost by allowing a building to remain functional after a natural disaster. It is likely that structural engineers will see a rising trend of using PBD, especially in buildings with expensive and complex architectural, electrical, mechanical, and plumbing systems. In fact, the 2010 edition of the ASCE 7 Standard, *Minimum Design Loads for Buildings and Other Structures*, has adopted quantitative performance-based design criteria in Chap. 1 as an alternative to the two traditional design philosophies, ASD and LRFD. And given that this standard is referenced by all major design codes, including the International Building Code, for structural design criteria and loadings, PBD should be easier to incorporate in the design of "green" buildings constructed in any material – concrete, steel, timber, masonry, or composites.

2.9 Conclusions

Through the use of sustainable practices in infrastructure design, the adverse environmental impact of new infrastructure systems can be minimized. More sustainable methods in the fabrication of construction materials (concrete, steel, timber, and masonry) can also drastically lower the overall construction cost, particularly when direct and indirect costs over the life of the system are considered; i.e., Life Cycle Assessment is the best approach to assess sustainability. Both Figs. 2.1 and 2.2 clearly show that without incorporating sustainability in construction, the adverse environmental effects from concrete and steel consumption would have continued to increase. If these effects are uncurbed, the environment will continue to suffer and CO_2 levels will keep increasing, continuing to threaten the long-term welfare and health of the public, which are within the purview of building codes.

Energy consumption is of more concern in cement (and concrete) production than the other construction materials. A temporary remedy for this issue is to incorporate more by-products (such as fly ash) in concrete; a more permanent solution is to find alternative carbon-neutral cementiceous materials, and ultimately lower the overall embodied energy in concrete. There are a number of researchers working on such a material, dubbed "green concrete." Steel is another construction material widely used in infrastructure. Its cradle-to-cradle property allows recycling to be done without affecting its performance, while at the same time reducing manufacturing costs and the impact to the environment. The timber industry has had the greatest impact on sustainability practices because the material can be considered a renewable resource. Furthermore, the industry has embraced sustainability practices at all levels, from harvesting to construction.

Both the concrete and steel industries have dramatically changed over time due to institutionalized groups like AISC, PCA, ACI, and LEED. Their collaborations with various other groups have allowed sustainability to be more accepted worldwide. These groups have redefined the role of engineers and have changed standards that have resulted in improved environmental policies. Current development of design specifications that incorporate more sustainable practices will allow sustainability to become common practice in future projects. In fact, a number of groups and government agencies have a goal of carbon-neutral materials and practices within the first part of the twenty-first century, see Fig. 2.4.

References

American Concrete Institute (ACI) (2002) Guide for the design and construction of externally bonded FRP systems for strengthening concrete structures. 440.2R-02. American Concrete Institute, Farmington Hills

American Institute of Steel Construction (AISC) (2011) The sustainable aspects of structural steel. Web, 26 Jan. 2011. http://www.aisc.org/uploadedFiles/Steel_Solutions_Center/Conceptual/My_Project/Files/Talking%20Points%20-%2010.2009.pdf

Arbuckle P (2010) Discovering institutional drivers and barriers to sustainable concrete construction. Charting courses toward sustainability, Center for Sustainable Systems. N.p., n.d. Web. 28 July 2010. http://css.snre.umich.edu/publication/discovering-institutional-drivers-and-barriers-sustainable-concrete-construction

ASTM C618-08a (2008) Standard specification for coal fly ash and raw or calcined natural pozzolan for use in concrete. ASTM International, West Conshohocken

ASTM C989-09a (2009) Standard specification for slag cement for use in concrete and mortars. ASTM International, West Conshohocken

Black S (2010) Life cycle assessment: are composites “green”? Composites Technology, Dec 2010

Bowyer J, Bratkovich S, Lindberg A, Fernholz K (2008) Wood products and carbon protocols: carbon storage and low energy intensity should be considered. Dovetail Partners, Inc., Minneapolis, 28 Apr 2008. www.dovetailinc.org

BRE, NetComposites (2004) Green guide to composites: an environmental profiling system for composite materials and products. BRE Bookshop, Garston

Brinson RA, Dolan JB (2008) Emerging risks of green construction; will LEED lead construction professionals to court more often? STRUCTURE Magazine. A joint publication of NCSEA/CASE/SEI, June 2008, Vol 15, No. 6

Calkins M (2009) Materials for sustainable sites, Har/Pas edn. Wiley, New York

DeStefano J (2009) Building green with wood construction. STRUCTURE Magazine. A joint publication of NCSEA/CASE/SEI, Aug 2009, Vol 16, No. 8

Falk RH (2010) Chapter 01: Wood as a sustainable building material. In: Wood handbook: wood as an engineering material, General technical report FPL-GTR-190. U.S. Department of Agriculture, Forest Service, Forest Products Laboratory, Madison, p 508

Hamburger RO (1997) A framework for performance-based earthquake resistive design. National Information Service for Earthquake Engineering. Web 6 Dec 2010. http://nisee.berkeley.edu/lessons/hamburger.html

Hays B, Cocke D (2009) Missed opportunities in structural engineering. STRUCTURE Magazine. A joint publication of NCSEA/CASE/SEI, Apr 2009, Vol 16, No. 4

Hewitt C (2003). The real deal: sustainable steel. Modern Steel Construction Magazine. American Institute of Steel Construction, Sept 2003

Hollaway LC (2010) A review of the present and future utilization of FRP composites in the civil infrastructure with reference to their important in-service properties. Constr Build Mater 24:2419–2445, Published by Elsevier Ltd

Kang GS, Kren A (2007) Structural engineering strategies towards sustainable design. SEAONC – Structural Engineers Association of Northern California. Web 31 Jan. 2011. http://www.seaonc.org/pdfs/SESTSD%20final%20%20.pdf

Kats GH (2003) Green building costs and financial benefits. http://www.nhphps.org/docs/documents/GreenBuildingspaper.pdf
Kestner DM, Goupil J, Lorenz E (2010) Sustainability guidelines for the structural engineer. ASCE, Reston
Kneer E, Maclise L (2008) Consideration of building performance in sustainable design: a structural engineer's role. In: Proceedings of the 2008 SEAOC (Structural Engineers Association of California) Convention, Hawaii, september 23–27, 15p
Lwin MM (2008) Sustainability considerations in bridge design, construction, and maintenance. Aspire Magazine, Winter 2008:49, Vol 2, No. 1
Mehta PK, Meryman H (2009) Tools for reducing carbon emissions due to cement consumption. STRUCTURE Magazine. A joint publication of NCSEA/CASE/SEI, Jan 2009, Vol 16, No. 1
Miller N, Spivey J, Florance A (2008) Does green pay off? J Real Estate Portf Manag, Oct 2008, Vol 14, No. 4, pp 385–399
Stubbles J (2000). Energy use in the U.S. steel industry: an historical perspective and future opportunities. Report under contract to Energetics, Inc. for the U.S. Department of Energy, Office of Industrial Technologies. Washington, DC
Sustainable-Steel.org (2010). Emission reduction. Sustainable steel: the EnviroMetal. Steel Recycling Institute, n.d. Web. 5 Aug 2010. http://www.sustainable-steel.org/emreduction.html
Tang M, Castro E, Pedroni F, Brzozowski A, Ettouney M (2008) Performance-based design with application to seismic hazard. STRUCTURE Magazine. A joint publication of NCSEA/CASE/SEI, June 2008, Vol 15, No. 6
Teller M, Bergman J (2010) Sustainable design for structural engineers. STRUCTURE Magazine. A joint publication of NCSEA/CASE/SEI, Dec 2010, Vol 17, No. 12
U.S. Environmental Protection Agency (EPA) (2006). Iron & steel. Web 4 Feb 2011. http://www.epa.gov/sustainableindustry/pdf/2006/ironandsteelbw.pdf
U.S. Environmental Protection Agency (EPA) (2009) Estimating 2003 building-related construction and demolition materials amounts. EPA530-R-09-002; Mar 2009
USGBC (2010) Greening the codes: building codes begin to broaden their charge to include human and environmental impacts of buildings into their health safety mission. US Green Building Council. Web. 31 Jan 2011. http://www.usgbc.org/
Volz V, Stovner E (2010a) Reducing embodied energy in masonry construction, Part 1: understanding embodied energy in masonry. STRUCTURE Magazine. A joint publication of NCSEA/CASE/SEI, May 2010, Vol 17, No. 5
Volz V, Stovner E (2010b) Reducing embodied energy in masonry construction, Part 2: evaluating new masonry materials. STRUCTURE Magazine. A joint publication of NCSEA/CASE/SEI, Sept 2010, Vol 17, No. 9
Ward R (2010) Can using more wood reduce your environmental footprint? STRUCTURE Magazine. A joint publication of NCSEA/CASE/SEI, Feb 2010, Vol 17, No. 2
Whittemore D (2010) Sustainable structures for the bridge engineer. STRUCTURE Magazine. A joint publication of NCSEA/CASE/SEI, Oct 2010, Vol 17, No. 10

Part I
Durability and Service Life

Chapter 3
Environmental Degradation of Interlaminar Shear Strength in Carbon/Epoxy Composites

Avinash Reddy Akepati, Abilash R. Nair, Samit Roy, Anwarul Haque, Piyush K. Dutta, and Ashok Kumar

Abstract The effect of environmental and loading conditions on the degradation of Interlaminar Shear Strength (ILSS) of the carbon-epoxy composite specimens was studied. The hygrothermal conditions capture the synergistic effects of field exposure and extreme temperatures. A short beam shear test (SBST) was performed to determine the Interlaminar Shear Strength (ILSS) of environmentally aged composite specimens in accordance with ASTM D2344-84. Initially, a standard two-dimensional cohesive layer constitutive model was employed in order to simulate the experiment using an in-house FEA code (NOVA-3D). Numerical instabilities, encountered using the standard cohesive layer model, were overcome by incorporating viscoelastic regularization in the constitutive equations of the cohesive layer. This modification also enabled the analysis to continue beyond the point of peak failure load. The model was able to accurately simulate the load vs. displacement behavior of most of the SBST samples aged under various hygrothermal and synergistically applied stress conditions. Further, the effect of displacement rate on the ILSS of specimens

A.R. Akepati (✉) • A.R. Nair • S. Roy • A. Haque
Department of Aerospace Engineering and Mechanics, University of Alabama,
209 Hardaway Hall, 7th Ave., Tuscaloosa, AL 35487, USA
e-mail: avinash@vt.edu; arnair@crimson.ua.edu; sroy@eng.ua.edu; ahaque@eng.ua.edu

P.K. Dutta
Dutta Technologies Inc., 4810 Euginea Drive, Palm Beach Gardens, FL 33418, USA
e-mail: pkdutta@aol.com

A. Kumar
U.S. Army Engineer Research and Development Center, Construction Engineering Research Laboratory (ERDC-CERL), Champaign, IL 61821-9005, USA
e-mail: Ashok.kumar@usace.army.mil

R. Jain and L. Lee (eds.), *Fiber Reinforced Polymer (FRP) Composites for Infrastructure Applications*, Strategies for Sustainability, DOI 10.1007/978-94-007-2357-3_3,

was studied using NOVA-3D. The model indicated a strong dependence of viscoelastic cohesive strength on the displacement rate. Regrettably, the predicted rate dependence could not be verified experimentally.

Keywords Carbon fiber reinforced polymers (CFRP) • Delamination • Interlaminar shear strength (ILSS) • Modeling • Viscoelastic cohesive layer

3.1 Introduction

Composites offer numerous advantages over conventional structural systems in the form of higher specific stiffness and strength and lower life-cycle costs with additional benefits, such as easier installation and improved safety. In recent years, there has been a considerable increase in the use of composites, especially carbon and glass fiber epoxy matrix composites, as appliqués and wraps for seismic upgrading and repairs. However, there is a general lack of data on the long-term performance of these materials. There are heightened concerns related to the overall durability of these materials, especially as related to their capacity for sustained performance under harsh and changing environmental conditions under static and seismic loads. They typically fail in a brittle and catastrophic manner with little warning. The current work aims to provide a means for predicting the long-term interlaminar shear strength of composite materials, based on laboratory testing and degradation modeling. Accelerated aging of the specimens was carried out in order to simulate various real world conditions. Laboratory experiments studied the degradation of ILSS of carbon/epoxy composites due to the synergistic influence of moisture, applied stress, and temperature. The degradation of ILSS for carbon/epoxy systems due to moisture absorption, is discussed in Zhuang and Wightman (1997) and Haque et al. (1991). Damage due to concurrently applied stress, moisture diffusion, and temperature plays a major role in composite degradation characteristics. Some studies report increased diffusion but unchanged concentration, while others report unchanged diffusion but increased saturation concentration (Patel and Case 2002).

Moisture absorption characteristics can be used to obtain insight into how and why the material is influenced by moisture. Zhuang and Wightman (1997) have reported interlaminar shear strength degradation for carbon/epoxy systems because of water absorption. They studied the effect of moisture on interfacial shear strength using single fiber fragmentation testing. Samples were stored in a 100% relative humidity environment at either 23°C or 75°C prior to testing. The results indicated that exposure to humidity does cause interfacial shear strength reduction. The strength reduction was found to be insensitive to the aging temperature. To distinguish between the effect of temperature and humidity, a parallel study was conducted on specimens aged at the same temperatures in dry conditions. No reduction in interfacial strength was found. The reduction in interfacial strength with humidity exposure was attributed to moisture-induced matrix plasticization and a subsequent decrease in matrix glass transition temperature (T_g).

The effect of isothermal aging on ILSS of quasi-isotropic composites laminates was evaluated as reported by Taylor and Lin (2003). Thermoset bismaleimide (IM7/5260) and thermoplastic polyimide (IM7/K3B) were investigated to determine the aging degradation in different matrix materials. Aging temperatures were around 150°C, 175°C, and 205°C, and aging time varied from a few hours to 28 months. Un-aged specimens were also tested for baseline data. Isothermal aging for IM7/5260 initially resulted in post-curing effects dominating, prior to an overall decrease in interlaminar shear strength due to thermal degradation. The interlaminar shear strength decreased for longer aging times, but showed little variation for different temperatures.

Delaminations in composites have been modeled using spring elements by Cui et al. (1993). The model was shown to be able to predict the onset and growth of the delamination by applying it to two types of delamination problems. One was a three point, short beam bending specimen, while the other was a cut-ply specimen. In both cases, the onset and growth of delamination was well predicted by the model. The model unified stress-based and fracture mechanics-based approaches for strength prediction. Also, the model required no assumption of the presence of initial defects.

FEA simulations of delamination were performed by Goruganthu et al. (2008) using the cohesive layer failure model to gain insight into the interlaminar failure mechanism of the composites after prolonged environmental exposure. But the model was unable to capture the load vs. displacement curve beyond peak load due to numerical instabilities.

One of the problems in modeling the failure of a weak interface using a cohesive-layer is the instability that is encountered due to elastic snap-back. The equations fail to converge at the point of instability due to an imbalance in strain energy. Introduction of a viscosity term in the equations of the cohesive layer helps in overcoming this problem through a "viscous regularization" of the response during unstable crack growth. In the present case, the "viscous regularization" approach is more suitable than the Riks Method (1979) because of its simplicity of implementation and numerical robustness. The viscoelasticity-based model was implemented in an in-house code (NOVA-3D) and successfully simulated catastrophic delaminations in short- beam shear experiments well beyond peak (failure) load.

3.2 Specimen Fabrication

Carbon fiber reinforced polymer (CFRP) specimens used for experimental testing were manufactured using Vacuum Assisted Resin Transfer Molding (VARTM). Unidirectional panels consisting of 30 zero-degree layers were fabricated. The polymer matrix used was SC-780, a two component toughened epoxy resin supplied by Applied Poleramic Inc., USA. The mixing ratio of resin to hardener was 4:1 by volume. The carbon fiber used was HMF 160T-650/35, supplied by SciArt, Canada. Figure 3.1 shows the setup of the manufacturing process.

Fig. 3.1 VARTM setup

3.3 Accelerated Aging of Short-Beam Shear Test Specimens

Two temperature conditions, termed 'hot' (H: 70°C) and 'cold' (C: 50°C), were selected for accelerated aging. Accelerated moisture aging of the specimens was carried out in environmental chambers at the two temperature conditions. In order to maintain 100% relative humidity (RH) over long periods of time, the 'hot wet' and 'cold wet' specimens were immersed in de-ionized water. Similarly, 'hot dry' specimens and 'cold dry' specimens were aged at 70°C for hot conditions and 50°C for cold conditions respectively. An applied load equivalent to 5% ultimate tensile strength (UTS) was applied to the "stressed" specimens during environmental aging using a unique constant-strain test fixture developed to avoid creep rupture of the specimens. Depending on the aging conditions, the specimens were designated as HWS, CWS, HWU, CWU, HDS, CDS, HDU, CDU (where, H: Hot, C:Cold, W:Wet, D:Dry, U:Unstressed, S:Stressed) followed by numbers 1, 2, or 3 representing the ageing time intervals t_1, t_2, and t_3.

3.4 Constant Strain Fixture

A unique constant strain fixture was employed in order to study the synergistic interaction between applied stress/strain and environmental conditions during ageing of test specimens. The design of the fixture aims to prevent premature rupture of the specimens due to tensile creep during aging. The fixture uses a beam and support rod mechanism in order to prevent excessive creep, as shown in Fig. 3.2. A total

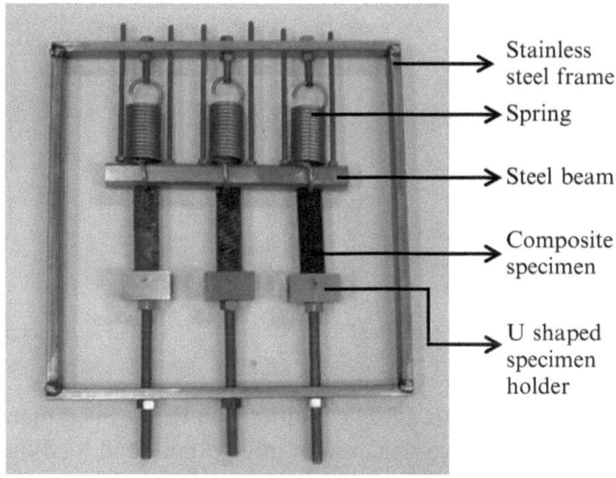

Fig. 3.2 Constant strain fixture used during aging

of 12 CFRP panels were aged using this fixture and premature failure was not observed in any of them during aging.

The test frame comprises of stainless steel bars welded together at the corners. Three springs are used to apply a constant continuous load. Selection of the springs depends on the amount of strain to be applied on the specimens and the deflection of the spring required from load measurement perspective. The deflection of the spring plays an important role in the design due to constraints on frame size. One end of the spring is fixed using a threaded rod and the other end of the spring is passed through a hole in the test specimen. Each specimen has a hole drilled at the other end and is fixed to the frame with a U-shaped specimen holder as shown in the Fig. 3.2. The holder is fixed to the frame with a threaded rod and nut. As the nut is turned clockwise, the rod travels downwards, thus elongating the specimen and extending the spring. The nut is turned continuously until the required spring deflection is achieved. The spring deflection is measured using a Vernier caliper. Once the frame is constructed, a steel beam is placed in the gap between the composite specimen and the spring hook, which passes through the specimen. The function of the beam is to restrain the specimen from further extending due to creep, thereby maintaining constant strain in the specimens, thus preventing premature specimen failure. The beam is held in place using threaded rods that are passed through the top frame and locked in place using nuts.

3.5 Mechanical Testing

Short-beam shear tests (SBST) were performed according to ASTM D2344-84 in order to determine the ILSS of composite specimens. SBST were performed on both aged and un-aged unidirectional CFRP specimens in order to investigate the

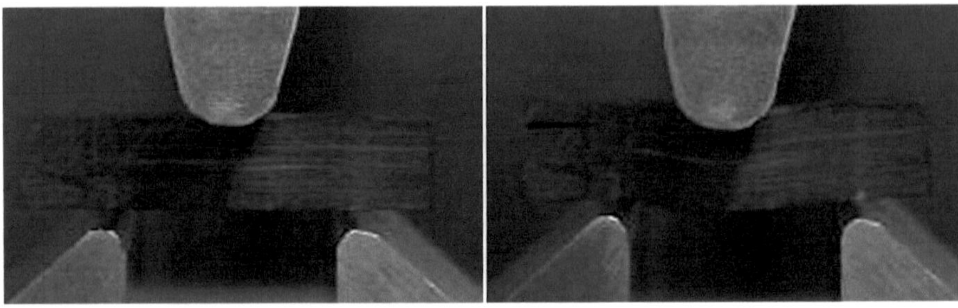

Fig. 3.3 Short-beam shear test specimen before and after delamination

effect of various aging conditions on ILSS. Specimens of small span-to-depth ratio (L/h = 3) were considered in three-point bend loading tests/experiments to produce horizontal shear failure between the laminates. The tests were performed using an MTS servo-hydraulic testing machine at room temperature and under ambient humidity. The rate of crosshead motion was maintained at 1.27 mm/min and load vs. displacement data was plotted. At least four replicate test specimens were tested for each ageing condition and for each ageing time. Figure 3.3 shows a short beam shear test specimen before and after delamination.

3.6 Modeling

An in-house NOVA-3D FEA code was used to simulate the short beam shear experiments conducted in the laboratory. Environment-assisted debonding along the interlaminar interface was simulated using a unique cohesive layer debonding model developed by Roy et al. (2006) that is capable of including environmental effects. Based on an approach discussed by Cui and Wisnom (1993) and extensively applied by Roy et al. (2006) and Roy and Wang (2005) to model debonding, a synergistic cohesive layer model was employed within the framework of finite element analysis (FEA) that effectively combines both stress-based and fracture-mechanics-based modeling for predicting interlaminar shear strength of CFRP.

As mentioned previously, numerical instabilities encountered during the SBST simulation using an elastic cohesive layer may be overcome using viscous regularization as proposed by Gao and Bower (2004). However, the viscous regularization is introduced in the form of artificial numerical damping, as shown in Gao and Bower (2004). It is now well established that resin rich areas in a composite laminate, such as interlaminar regions, exhibit viscoelastic behavior at elevated temperatures and high relative humidity under translaminar loading by Crossman et al. (1978). Employing a micromechanics-based viscoelastic traction separation behavior proposed by Allen and Searcy (2001), it is herein proposed that the sudden release of elastic strain energy at the time of failure, which causes instability, is dissipated using natural viscoelastic damping in the resin rich interlaminar region. Consequently, the instability encountered at failure initiation with elastic cohesive elements is circumvented; furthermore, as an added bonus, it enables tracking of the nonlinear load vs. displacement curve well beyond peak load.

3.7 Multi-axial Viscoelastic Cohesive Layer Derivation Including Damage Evolution

From Allen and Searcy (2001), the area averaged fibril tractions across the cross-sectional area of the cohesive layer representative volume element (RVE), as shown in Figs. 3.4 and 3.5, may be expressed by,

$$T_i = \frac{1}{A}\int_A t_i^{\text{fibril}} dA; \text{ i} = 1,2,3 \tag{3.1}$$

where, A is the cross-sectional area of the RVE, and T_i is the ith rectangular Cartesian component of traction. Assuming uniform traction within each fibril of the RVE, Eq. 3.1 can be discretized as,

$$T_i = \sum_{k=1}^{N} \frac{A_k(t)}{A} t_i^{\text{fibril}}; \text{ i} = 1,2,3 \tag{3.2}$$

where, A_k represents the cross-sectional area of the kth fibril within the RVE, and N is the number of fibrils in the RVE. From continuum mechanics, components of surface traction vector at the mid-plane of each fibril can be related to the components of the Cauchy stress tensor by

$$\sigma_{ij}^{\text{fibril}} n_j = t_i^{\text{fibril}} \tag{3.3a}$$

$$\bar{\sigma}_{ij} n_j = T_i \tag{3.3b}$$

where the overbar represents area averaged quantity (average stress) and n_j are the rectangular Cartesian components of the direction cosines at the fibril mid-plane.

Substituting Eqs. 3.3a and 3.3b in Eq. 3.2

$$\bar{\sigma}_{ij} n_j = \sum_{k=1}^{N} \frac{A_k(t)}{A} \sigma_{ij}^{\text{fibril}} n_j$$

$$\left\{\bar{\sigma}_{ij} - \sum_{k=1}^{N} \frac{A_k(t)}{A} \sigma_{ij}^{\text{fibril}}\right\} n_j = 0 \tag{3.4}$$

Since Eq. 3.4 must hold for any arbitrary orientation of the normal to the fibril mid-plane; therefore

$$\bar{\sigma}_{ij} = \sum_{k=1}^{N} \frac{A_k(t)}{A} \sigma_{ij}^{\text{fibri}} \tag{3.5}$$

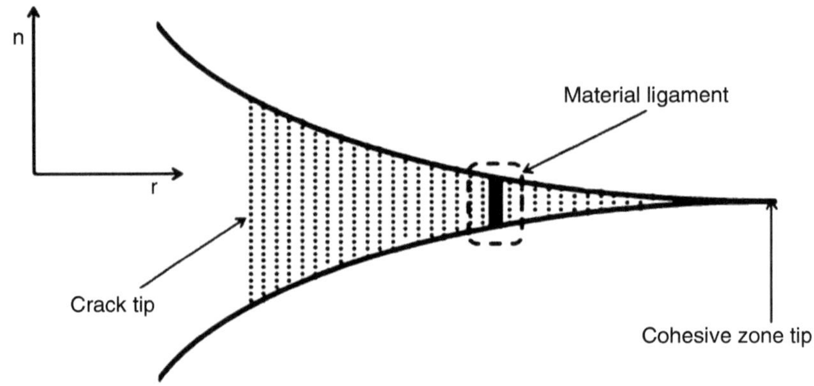

Fig. 3.4 Opening crack containing cohesive ligament

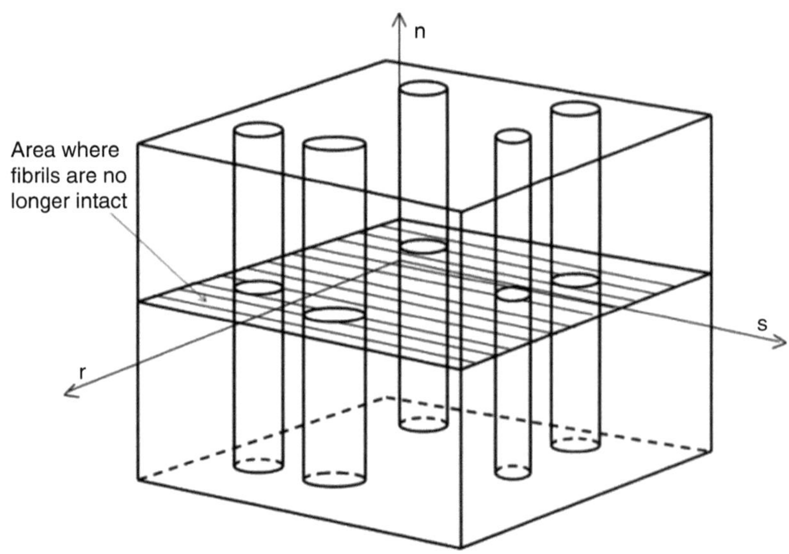

Fig. 3.5 Reduction of RVE to cohesive zone by area averaging fibril tractions

Defining a continuum internal damage parameter α representing the time varying area fraction of the growing voids with respect to the cross-sectional area of the RVE

$$\alpha(t) = \frac{A - \sum_{k=1}^{N} A_k(t)}{A} \tag{3.6}$$

Consequently, Eq. 3.5 reduces to

$$\bar{\sigma}_{ij} = (1-\alpha(t))\,\sigma_{ij}^{\text{fibril}} \tag{3.7}$$

As described in Roy and Reddy (1988), the multi-axial viscoelastic stress-strain law for a polymer fibril may be expressed in matrix notation as

$$\{\sigma(t)\}^{\text{fibril}} = [M(t)](\{\varepsilon(t)\}-\{H(t)\}) \tag{3.8}$$

where [M(t)] is a 6×6 matrix of time-dependent viscoelastic stiffness components, $\{\varepsilon(t)\}$ is the vector containing the components of mechanical strains at time t, and $\{H(t)\}$ contains the hereditary (load history-dependent) strain components. Details regarding the derivation of the [M] matrix and {H} vector from the viscoelastic convolution integral can be found in Roy and Reddy (1988). Combining Eqs. 3.7 and 3.8 gives the constitutive relationship between viscoleastic stresses and strains within a cohesive RVE at an interlaminar interface, including evolving damage

$$\{\bar{\sigma}(t)\} = (1-\alpha(t))[M(t)](\{\varepsilon(t)\}-\{H(t)\}) \tag{3.9}$$

3.8 Damage Evolution Law

In the absence of test data regarding the change of fibril diameter as a function of time, a phenomenological damage evolution law is adapted, given by

$$\frac{d\alpha}{dt} = \begin{cases} \alpha_0\bar{\lambda}^m, \text{ if } \dot{\lambda} \geq 0 \text{ and } \alpha < 1 \\ 0, \text{ if } \dot{\lambda} < 0 \text{ and } \alpha < 1 \end{cases} \tag{3.10}$$

where, $\bar{\lambda}$ is a principal stretch measure within the RVE (refer to Eq. 3.18 for details), and α_0 and m are material constants assumed to be independent of environmental conditions.

3.9 Determination of Principal Stretch

At any point within the RVE, the deformation gradient $F_{ij}(t)$ is given by

$$F_{ij}(t) = \frac{\partial x_i}{\partial X_j} \tag{3.11}$$

Where x_i is the current configuration and X_j is the reference configuration. The right Cauchy-Green tensor C_{ij} is given by (Malvern 1969)

$$C_{ij} = \frac{\partial x_k}{\partial X_i}\frac{\partial x_k}{\partial X_j} \tag{3.12}$$

In matrix notation, Eq. 3.12 becomes

$$[C] = [F]^T[F] \tag{3.13}$$

Assuming a multiplicative decomposition of the deformation gradient into stretch [U] and a rotation [R]

$$[F] = [R][U] \tag{3.14}$$

Equation 3.13 becomes

$$[C] = [U]^T[R]^T[R][U] = [U]^T[U] = [U]^2 \tag{3.15}$$

Therefore, the principal eigen value (λ_I) of the stretch tensor [U] can be obtained by taking positive square-root of the highest principal eigen value of [C]

$$U_I = \sqrt{C_I} = \lambda_I \tag{3.16}$$

3.10 Damage Initiation Criterion

Within the cohesive RVE, damage initiation is assumed to occur if a critical value of the Von-Mises equivalent stress $\sigma_{M,cr}$ is exceeded (at each element Gauss point), i.e.

$$\sigma_{M,cr} = \sqrt{\frac{\left[(\sigma_{xx}-\sigma_{yy})^2+(\sigma_{yy}-\sigma_{zz})^2+(\sigma_{xx}-\sigma_{zz})^2+6(\tau_{xy}^2+\tau_{xz}^2+\tau_{yz}^2)\right]}{2}} \tag{3.17}$$

If Eq. 3.17 is satisfied, then Eq. 3.10 is invoked such that

$$\bar{\lambda}(t) = \lambda_I(t) - \lambda_{CR}, \quad \lambda_I(t) \geq \lambda_{CR} \tag{3.18}$$

where $\lambda_I(t)$ is the current value of the principal stretch at time t, and λ_{CR} is the critical value of principal stretch at damage initiation.

The Prony series coefficients for the creep compliance of the resin required as input to the FEA model were obtained from the work of Muliana and Haj-Ali (2003). They developed a micromechanical model to derive the nonlinear viscoelastic response of laminated composites. Glass/epoxy composite specimens of various

Table 3.1 Damage parameters used in FEA

Parameter	Value
α_0	0.032
m	0.750

Table 3.2 Values of critical von Mises stress used in FEA for different aging conditions

At $t=t_3$ (approximately 5,000 h)		
	Critical von Mises stress (MPa)	Percentage change
Control specimen	119	0.00
HWU	63	–47.06
CWU	64	–46.22
HWS	50	–57.98
CWS	70	–41.18
HDS	123	3.36
CDS	100	−15.97
HDU	130	9.24
CDU	135	13.44

off-axis angles were used to calibrate their model and the Prony series coefficients were obtained for the epoxy matrix used in their work. The value of various damage parameters and critical Von Mises stress used in the FEA model are listed in Tables 3.1 and 3.2 respectively.

3.11 Results and Discussion

3.11.1 Experimental Test Data

Short beam shear tests were initially performed on control specimens followed by the testing of various aged specimens at three aging time intervals t_1, t_2, and t_3. The aging time for each interval for various aging conditions, along with the number of aged specimens, is shown in Table 3.3. The experimental results of SBST on the control specimen are shown in Fig. 3.6 and Table 3.4. A total of five specimens were tested and the average ILSS was determined to be 50.42 MPa, as shown in Table 3.4.

The ILSS for various aged specimens at the three time intervals is shown in Table 3.5. A displacement rate of 1.27 mm/min was used in all experiments. Figure 3.6 shows the load vs. displacement curves for the five control specimens.

Figure 3.7 shows the variation of ILSS for various aged specimens with time. Figure 3.7 shows that degradation in ILSS occurs primarily at the first time interval t_1, which is the time taken for the wet specimens to saturate with moisture absorption. Thereafter, the ILSS remains more or less constant. Another important observation is that the observed strength of the dry specimens is much higher compared with

Table 3.3 Number of specimens aged for SBS test and their aging times

Environmental exposure condition	Number of specimens [ageing time (hours)]				
	t_0	t_1	t_2	t_3	Total
HWU	–	4 (1,684)	4 (3,378)	4 (4,883)	12
HDU	–	4 (1,968)	4 (3,768)	4 (5,448)	12
CWU	–	4 (1,680)	4 (3,333)	4 (4,837)	12
CDU	–	4 (1,920)	4 (3,768)	4 (5,448)	12
HWS	–	4 (1,016)	4 (2,546)	4 (4,128)	12
HDS	–	4 (1,992)	4 (3,504)	4 (5,016)	12
CWS	–	4 (1,016)	4 (2,616)	4 (4,198)	12
CDS	–	4 (1,920)	4 (3,432)	4 (4,944)	12

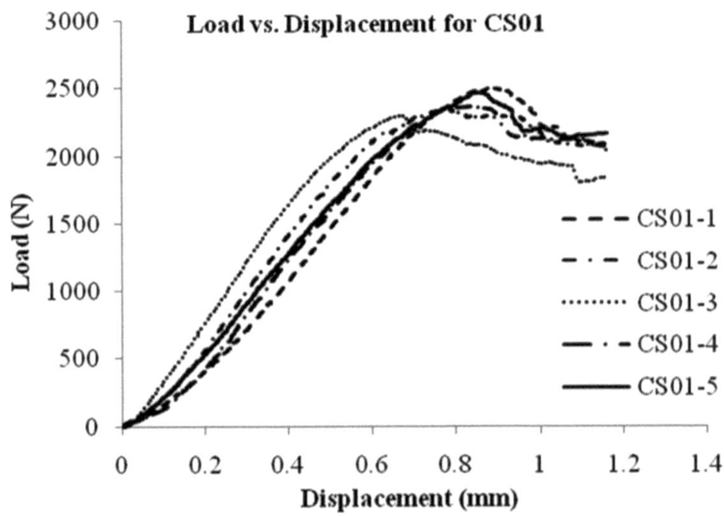

Fig. 3.6 Load vs. displacement curves for the control sample

the wet ones. This clearly establishes moisture as the primary cause of degradation in CFRP.

The average ILSS of dry specimens is 50.70 MPa, which is 0.55% higher compared to the 50.42 MPa of the control specimens. Hence, for the dry case, neither temperature nor applied stress seems to affect the strength of the composite. There are some oscillations in ILSS between the aging time intervals, likely due to variability between specimens. The slight increase in the strength of some dry specimens is possibly due to the post-curing effect. In case of wet specimens, the average ILSS is 38.88 MPa, a decrease of 23% compared with control specimens. Under wet conditions, temperature does not seem to be driving the degradation. Stress, when applied in conjunction with moisture, produces a faster rate of degradation, as can be seen from the HWS and CWS curves in Fig. 3.7. Thus, moisture plays a very significant role in the degradation of ILSS. Further, a synergistic combination of moisture and stress results in the faster rate of degradation compared to the unstressed case.

Table 3.4 Interlaminar shear strength data for control specimens

Number	Specimen	τ_{max} (MPa)
1	CS01-1	54.46
2	CS01-2	49.54
3	CS01-3	49.22
4	CS01-4	48.42
5	CS01-5	50.48
Average		**50.42**
Standard deviation		**±2.37**

Table 3.5 ILSS of various specimens at the three aging times

	Interlaminar shear strength (ILSS) (MPa)							
	50.42 (baseline)							
CS01	HWU	CWU	HDU	CDU	HWS	CWS	HDS	CDS
t_1	37.76	36.75	49.89	45.56	37.60	37.82	55.04	47.68
Standard deviation	±1.08	±3.04	±1.54	±1.75	±0.26	±1.20	±2.00	±0.96
Percentage change	**−25.10**	**−27.11**	**−1.05**	**−9.63**	**−25.42**	**−25.00**	**9.16**	**−5.43**
t_2	36.57	37.83	51.28	50.53	39.34	40.81	53.11	52.80
Standard deviation	±1.27	±1.48	±0.44	±1.80	±1.95	±1.12	±2.84	±2.61
Percentage change	**−27.46**	**−24.97**	**1.70**	**0.22**	**−21.97**	**−19.05**	**5.33**	**4.72**
t_3	37.99	39.09	50.83	50.75	37.74	40.71	53.79	47.46
Standard deviation	±1.09	±1.96	±1.35	±2.17	±0.92	±2.72	±2.93	±1.83
Percentage change	**−24.64**	**−22.46**	**0.81**	**0.65**	**−25.68**	**−19.25**	**6.68**	**−5.87**

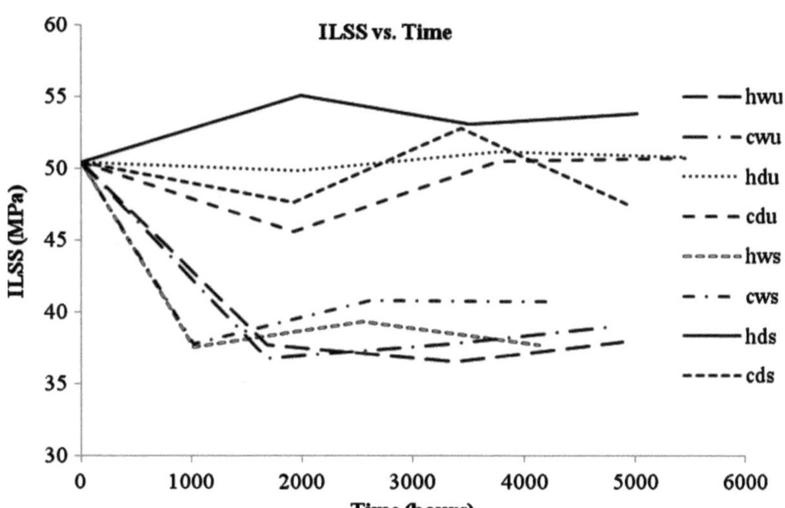

Fig. 3.7 Change in ILSS vs. time

3.12 FEA Simulation of Experiments

A two-dimensional FEA model was used to simulate the SBST experiments. A 2-D mesh was generated as shown in Fig. 3.8. The mesh consisted of a total of 3,354 elements, out of which 234 elements were viscoelastic cohesive elements. The size of the mesh was determined based on a convergence analysis. Three layers of viscoelastic cohesive elements, along the length of the specimen, were used in the mesh to simulate delaminations. Figure 3.9 shows a zoomed-in view of the viscoelastic cohesive layers. The location of the three cohesive layers is indicated by 'V' in Fig. 3.9.

The FEA model accurately simulated the actual SBST experiment for most of the aging conditions. The analysis was first carried on the control specimen and then extended to various aged specimens. The effect of thickness of the cohesive layer on solution convergence was also studied. As long as the same aspect ratio is maintained, the thickness of the element cohesive layer does not affect the predicted peak load by more than 5%. The deformation contour plot in Fig. 3.10 shows the delamination in the specimen. A zoomed-in view of the delamination in the cohesive layer is shown in Fig. 3.11. Figure 3.12a shows the experimental and simulated load vs. displacement curves for the control sample, and Fig. 3.12b shows the experimental and simulated load vs. displacement curves for all the aging conditions for time t_3. For all cases, error in peak load prediction was less than 10%.

The viscoelastic cohesive layer used in the model helped in attaining numerical stability at failure and the analysis could continue beyond the point of failure. Thus, the load vs. displacement curve was tracked well beyond peak load, without recourse to Riks algorithm (1979).

3.13 Effect of Displacement Rate

FEA simulations were carried out using NOVA-3D for the SBST experiments by varying the displacement rate. HWS3 aging condition was considered for this purpose. The experimental displacement rate of 1.27 mm/min was considered as baseline. The analysis was done at displacement rates of 0.254 mm/min and 2.54 mm/

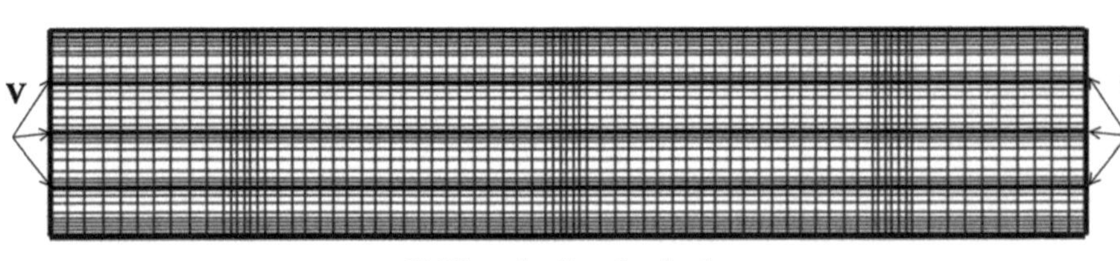

Fig. 3.8 Mesh used for FEA analysis

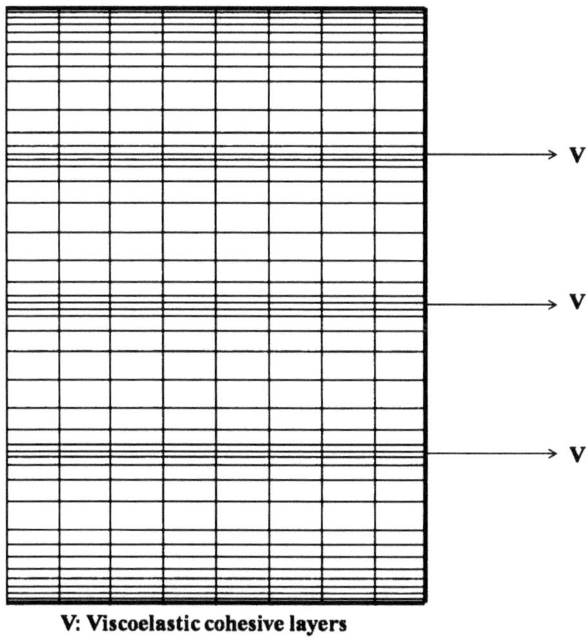

Fig. 3.9 Zoomed-in view of the cohesive layers

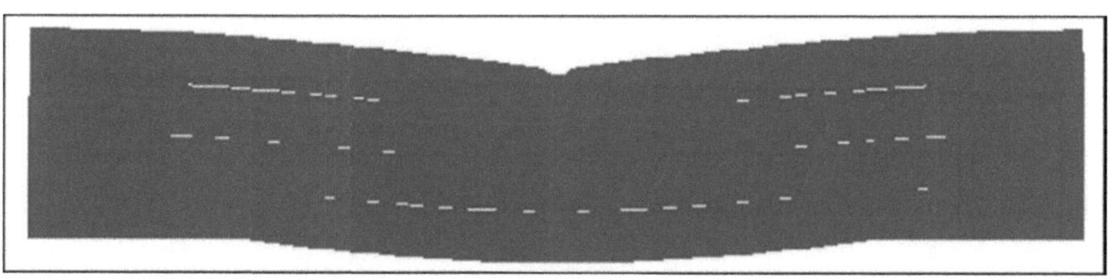

Fig. 3.10 Crack propagation in SBST simulation

min while keeping all other parameters constant. As one might expect for a rate dependent epoxy polymer, the FEA simulations indicate a strong dependence of viscoelastic cohesive strength on the displacement rate, as shown in Fig. 3.13. At lower displacement rates, the specimen fails at a lower load and the failure load increases as the displacement rate increases. Regrettably, the predicted rate dependence could not be verified experimentally because of premature failure due to high void content in the test specimens used in the elevated, as well as decreased strain rate experiments.

Figure 3.14 shows the predicted stress vs. strain plot within a cohesive element in the path of the delamination for two aging conditions (wet and dry). It can be observed that the wet sample has significantly lower strength and fracture toughness (indicted by the area under the curve) compared with the dry one. Figure 3.15 shows the predicted stress-strain response for various displacement rates within a cohesive layer element for a particular aging condition. The predicted rate dependence for the cohesive traction separation law is similar to the one for the case of load vs. displacement plots in Fig. 3.13.

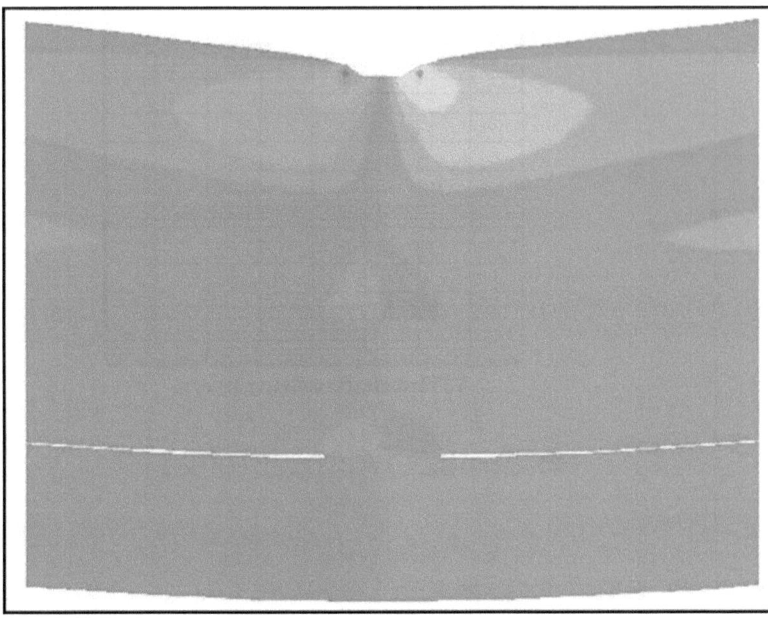

Fig. 3.11 Zoomed-in view of the delamination in cohesive layer

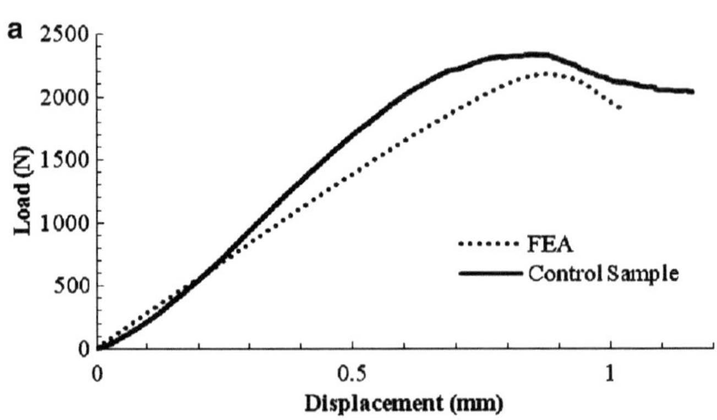

Fig. 3.12 (**a**) Experimental and FEA load vs. displacement curves for control sample for SBST. (**b**) Comparison of experimental and simulation results for various aging conditions for t_3

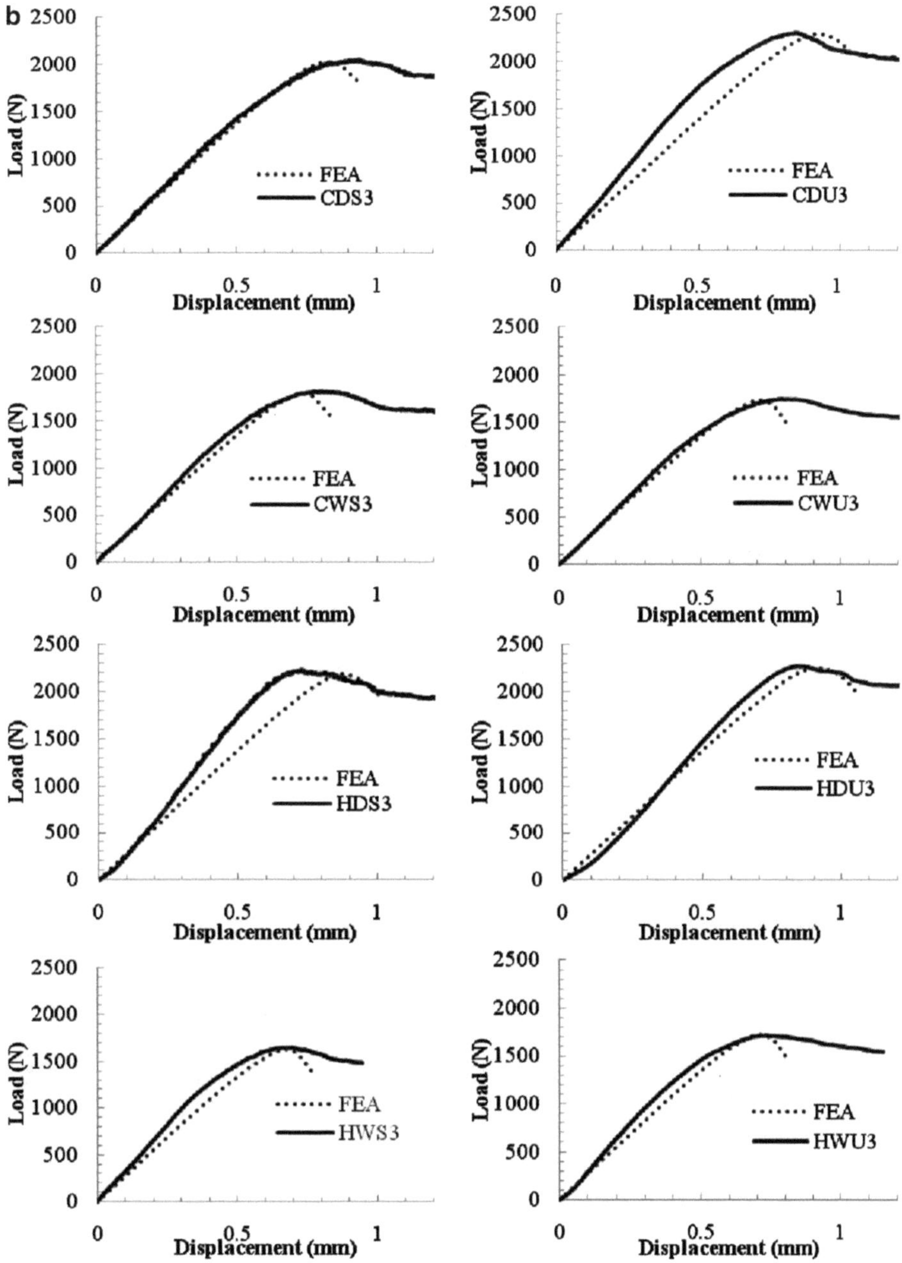

Fig. 3.12 (continued)

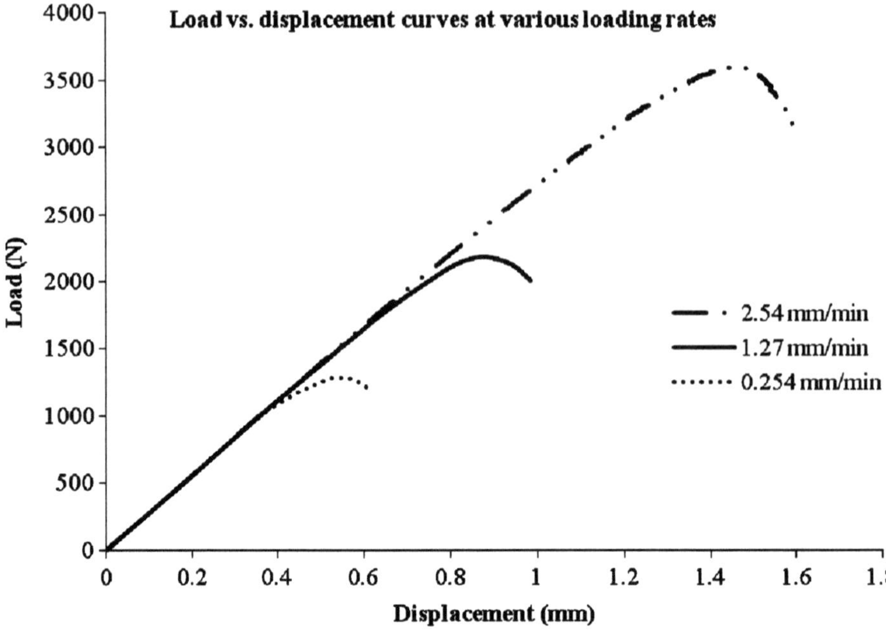

Fig. 3.13 Load vs. displacement plot for various displacement rates

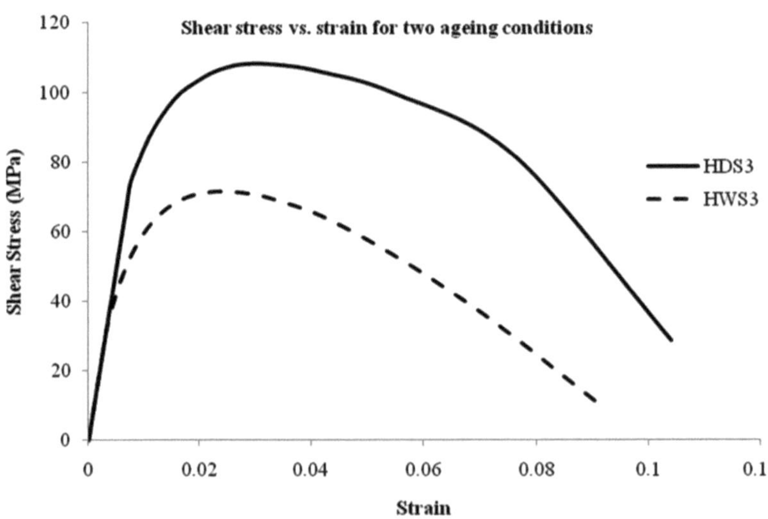

Fig. 3.14 Stress vs. strain response for HDS3 and HWS3

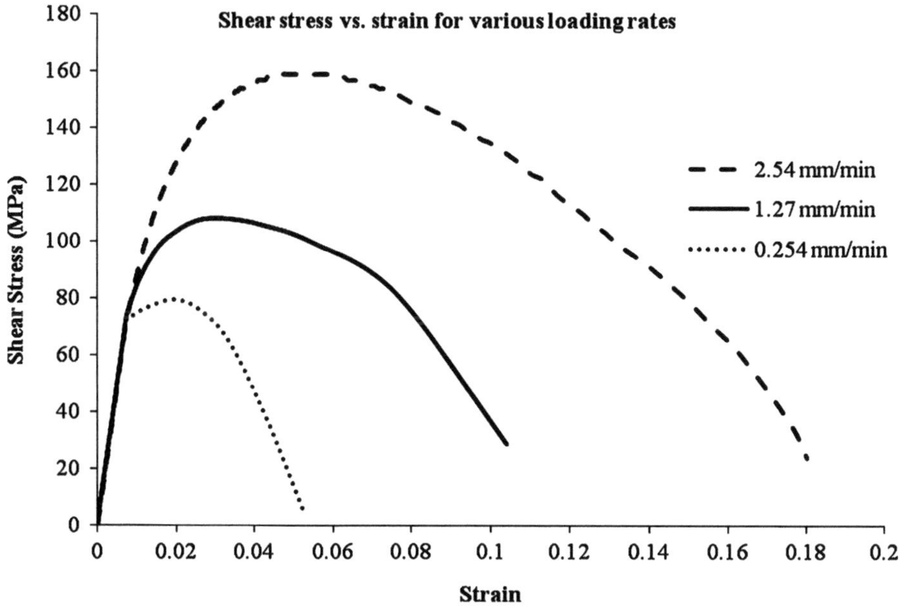

Fig. 3.15 Stress v/s strain response for various displacement rates

The results obtained in the simulation are compared with the results of an earlier analysis using purely elastic cohesive layer, as shown in Fig. 3.16. Although the materials used and the traction-separation law used was different, the comparison clearly shows the ability of viscoelastic cohesive layer to continue the analysis beyond peak load due to the mitigating effect of viscous regularization, which was not the case with the earlier elastic analysis.

3.14 Conclusions

This paper describes the synergistic effect of environmental conditions and applied loading on interlaminar shear strength of carbon/epoxy composites. The SBST test results indicate that ILSS of hot/dry specimens increased slightly due to post cure, whereas that of the hot/wet specimens significantly decreased by almost 23% after prolonged environmental exposure. Therefore, it can be concluded that moisture plays a very significant role in the degradation of ILSS. A synergistic combination of moisture and stress results in the faster rate of degradation compared to the unstressed case. ILSS appears to reach a threshold value after about 3,500 h of aging time. The experimental results of the short-beam shear test were simulated using viscoelastic cohesive layer-based finite element analysis. The analysis was found to be in good agreement with the experiment for various aging conditions. This study

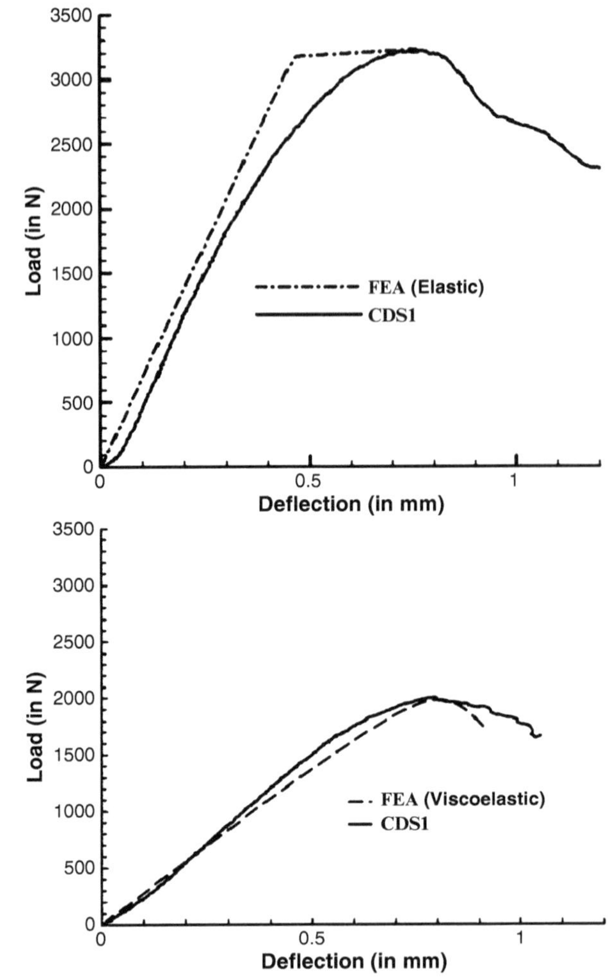

Fig. 3.16 Comparison of load vs. displacement curves of an elastic cohesive layer model (from earlier analysis) and viscoelastic cohesive layer model for CDS1

demonstrates the ability of a viscoelastic cohesive layer to continue the FEA analysis beyond peak load due to the benign effect of viscous regularization, which was not the case with earlier elastic analysis.

Acknowledgement The authors would like to acknowledge the support of this work by the Construction Engineering Research Laboratory (CERL), US Army Engineer Research and Development Center (ERDC), under Army contract W9132T07C0025/STTR-Phase II.

References

Allen DH, Searcy CR (2001) A micromechanical model for a viscoelastic cohesive zone. Int J Fract 107:159–176

Crossman FW, Mauri RE, Warren WJ (1978) Moisture altered viscoelastic response of graphite/epoxy composite. Adv Compos Mater Environ Effects ASTM STP 658:205–220

Cui W, Wisnom MR (1993) A combined stress-based and fracture-mechanics-based model for predicting delamination in composites. Composites 24:467–474

Gao YF, Bower AF (2004) A simple technique for avoiding convergence problems in finite element simulations of crack nucleation and growth on cohesive interfaces. Model Simul Mater Sci Eng 12:453–463

Goruganthu S, Elwell J, Ramasetty A, Nair AR, Roy S, Haque A, Dutta PK, Kumar A (2008) Characterization and modeling of the effect of environmental degradation on interlaminar shear strength of carbon/epoxy composites. Polym Polym Compos 16:165–179

Haj-Ali RM, Muliana AH (2003) A micromechanical model for the nonlinear viscoelastic behavior of laminated composites. Int J Solids Struct 40:1037–1057

Haque A, Mahmood S, Walker L, Jeelani S (1991) Moisture and temperature induced degradation in tensile properties of Kevlar-Graphite/epoxy hybrid composites. J Reinf Plast Compos 10:132–145

Malvern LE (1969) Introduction to the mechanics of continuous medium. Prentice-Hall, Englewood Cliffs

Patel SR, Case SW (2002) Durability of hygrothermally aged graphite/epoxy woven composite under combined hygrothermal conditions. Int J Fatigue 24:1295–1301

Riks E (1979) An incremental approach to the solution of snapping and buckling problems. Int J Solids Struct 15:529–551

Roy S, Reddy JN (1988) Finite-element models of viscoelasticity and diffusion in adhesively bonded joints. Int J Numer Methods Eng 26(11):2531–2546

Roy S, Wang Y (2005) Analytical solution for cohesive layer model and model verification. Polym Polym Compos 13(8):741–752

Roy S, Wang Y, Park LKM (2006) Cohesive layer modeling of time-dependent debond growth in aggressive environments. J Eng Mater Technol 128:11–17

Taylor DM, Lin KY (2003) Aging effects on the interlaminar shear strength of high-performance composites. J Aircr 40(5):971–976

Zhuang H, Wightman JP (1997) Influence of surface properties on carbon fiber/epoxy matrix interfacial adhesion. J Adhes 62(1–4):213–245

Chapter 4
Predicting Hygrothermal Degradation of Composites in Accelerated Testing

Jonathan Trovillion

Abstract Advanced fiber reinforced polymer (FRP) composite materials have been increasingly used in many applications relevant to the Army's transformation. Many of these applications require the FRP composites to perform over long periods in harsh environments with extremes of temperature, humidity, water, and exposure to ultraviolet radiation and chemicals. It is important to understand the long term durability of FRP composites to environmental stimuli.

This paper presents results of the hygrothermal degradation of E-glass/epoxy composites in accelerated tests and compares these results to predictions made using a modeling methodology based on Arrhenius-type reaction laws. To investigate the hygrothermal degradation behavior, E-glass/epoxy composites were subjected to accelerated tests at controlled temperatures and relative humidities. The specimens were exposed in an unloaded state and with a static tensile load of 2% of the ultimate transverse tensile strength. In the model predictions, three degradation mechanisms were considered: (1) post-curing, (2) thermal degradation, and (3) hygrothermal degradation.

Keywords Fiber reinforced polymer (FRP) composites • Composites • Degradation • Hygrothermal • Durability

4.1 Introduction

Advanced fiber reinforced polymer (FRP) composite materials have been increasingly used in many applications relevant to the Army's transformation. In general, FRP composites offer high stiffness and strength-to-weight ratios and are durable

J. Trovillion (✉)
US Army Engineer Research and Development Center – Construction Engineering Research Laboratory (ERDC-CERL), PO Box 9005, Champaign, IL 61826, USA
e-mail: jonathan.c.trovillion@usace.army.mil

R. Jain and L. Lee (eds.), *Fiber Reinforced Polymer (FRP) Composites for Infrastructure Applications*, Strategies for Sustainability, DOI 10.1007/978-94-007-2357-3_4,

in harsh environments. To date, the Army has researched FRP composites for applications ranging from the seismic rehabilitation of structures, high-performance tank armor, composite skins in planes and helicopters, ballistic panels, rotorcraft blades, and ballistic protective inserts (BPI) in personnel armor. Many of these applications require the FRP composites to perform over long time periods in harsh environments with extremes of temperature, humidity, water, and exposure to ultraviolet radiation and chemicals. It is important to understand the long term durability of FRP composites to environmental stimuli in order to effectively use FRP composites.

The objective of this research is to investigate degradation mechanisms and develop degradation models for prediction of long term service life of FRP composite materials for use in Army facilities. To investigate the hygrothermal degradation behavior, E-glass/epoxy composites were subjected to accelerated tests at controlled temperatures and relative humidities using a temperature humidity cabinet. The composite specimens were initially exposed to two temperatures (40°C and 85°C) and three relative humidities (93%, 50%, and 2%). The specimens were also exposed in an unloaded state, and with a static tensile load of 2% of the ultimate transverse tensile strength. The specimens were removed at regular intervals and tested for interlaminar (short-beam-shear) shear strength. Additional tests were conducted at temperatures above 100°C, without humidity control, to separate the thermal degradation mechanism.

This paper presents the specifics of the accelerated test procedures, including the FRP composite panel manufacture, specimen preparation, exposure environments, and experimental procedures and results. A prediction modeling methodology will be introduced and used to predict the hygrothermal degradation of unidirectional and quasi-isotropic composite laminates.

4.2 Experimental Program

To study the degradation of composites, it is important to manufacture specimens with a high degree of quality control. Toward this end, the E-glass epoxy composite plates were manufactured using a resin infusion molding (RIM) technique. The composite plates were manufactured using Hexcel unidirectional E-glass cloth (95% in fiber direction, 5% perpendicular to fiber direction) 6 oz per square yard (purchased from Fiberglassite) and SC-780 toughened epoxy resin from Applied Poleramic, Inc. Resin infusion was accomplished under full vacuum and held 24 h, after which the plates were removed from the vacuum bag. No post cure was used; rather, the specimens remained in an indoor environment at room temperature until they were cut into smaller specimens. Panels with lay-ups of $[0]_{10}$ and $[0]_{42}$ unidirectional, and $[\pm 45/0/90]_s$ quasi-isotropic fiber orientations were fabricated. The unidirectional panels were 0.3 m by 0.95 m (12 in. by 37.5 in.) by 6.4 mm (0.25 in.) thick for the $[0]_{42}$ and 1.5 mm (0.06 in.) for the $[0]_{10}$. The quasi-isotropic panels were 0.46 m by 0.46 m (18 in. by 18 in.) and 1.2 mm (0.05) thick.

The composite plates manufactured with the RIM technique were cut into 0.28 m by 0.025 m (11 in. by 1 in.) coupons using a water jet cutting technique. Two small holes were drilled at the ends of the coupons to facilitate connection to the exposure racks. After cutting and drilling, the specimens were wiped clean and placed in desiccators to remove any residual moisture from the samples. After desiccation, the specimens were mounted on special racks and placed in the temperature/humidity cabinet.

The $[0]_{42}$ unidirectional composite samples were exposed to selected temperature and relative humidity conditions using a Cincinnati Sub Zero temperature and humidity chamber Model Z32 Plus. This chamber has an interior dimensions of $0.97\times0.97\times0.97$ m and several horizontal racks. The composite specimens were initially exposed to two temperatures (40°C and 85°C) and three relative humidities (93%, 50% and 2%). The specimens were exposed in an unstressed state, and with a static tensile stress of 2% of the ultimate transverse tensile strength. The specimens were removed at regular intervals and tested for interlaminar (short-beam-shear) shear strength. A second series of interlaminar shear tests were conducted at temperatures of 100°C, 125°C, and 150°C, without humidity control in order to separate the thermal degradation mechanism.

A separate series of tests were conducted to provide data to compare against the model predictions. These included panels with lay-ups of $[0]_{10}$ and $[0]_{42}$ unidirectional, and $[\pm45/0/90]_s$ quasi-isotropic. The $[0]_{42}$ unidirectional were tested in interlaminar shear and the $[0]_{10}$ unidirectional and $[\pm45/0/90]_s$ quasi-isotropic were tested in tension. These were exposed at two temperature and humidity conditions: 85°C at 50% relative humidity, and 125°C and dry (no humidity control).

Changes in the material properties of the composites were measured using both short beam shear and tensile tests. The interlaminar shear tests were conducted according to ASTM D2344 (2006) in a fixture designed to conduct these tests. For these tests, the maximum applied load was recorded and the maximum interlaminar shear strength was calculated. The tensile tests were conducted according to ASTM D3039 (2010). A biaxial rosette strain gage was applied to the tensile coupons to record axial and transverse strain during testing. The tensile modulus and ultimate strength were calculated for each test. Both tensile and short beam shear tests were conducted in a screw driven United Test Machine.

4.3 Experimental Results

The degradation data for the short beam shear tests from the 6 exposure conditions are given in Table 4.1. This data is presented in the form of percent reduction of short beam shear strength compared to the baseline value as a function of time. Table 4.1 gives the experimental results for the specimens that stressed to 2% of the ultimate transverse tensile strength during exposure.

The data for the 40°C stressed cases at relative humidities of 2%, 50%, and 93% are plotted in Fig. 4.1. The data plotted is the average value of a series of five tests,

Table 4.1 Change in interlaminar shear strength for specimens exposed at 40°C and 85°C

Exposure time (weeks)		2% Relative humidity	50% Relative humidity	93% Relative humidity
Change in interlaminar shear strength for specimens exposed at 40°C and stressed				
0	Baseline value (MPa)	51.0	42.0	51.0
2	Percent change	10.6	24.2	11.4
4	Percent change	−13.0	11.8	−17.9
6	Percent change	−1.66	14.8	−30.8
Change in interlaminar shear strength for specimens exposed at 85°C and stressed				
0	Baseline value (MPa)	42.0	44.4	45.0
2	Percent change	26.2	6.80	−29.42
4	Percent change	35.1	–	−34.21
6	Percent change	41.9	6.30	–

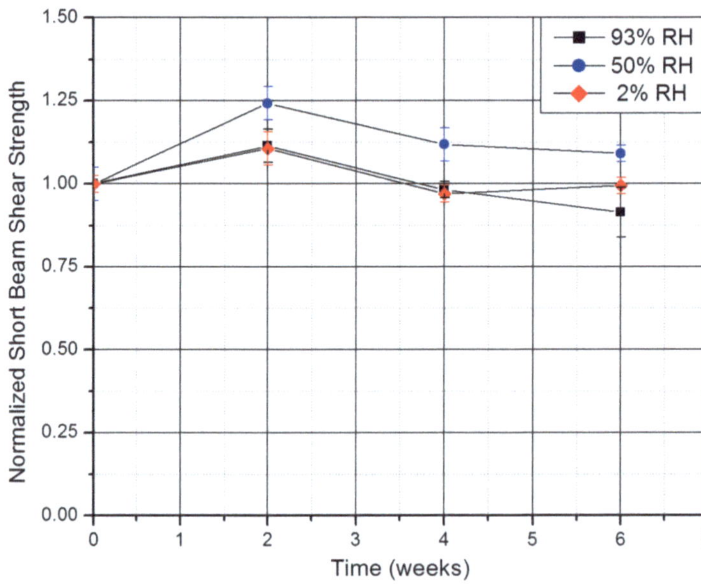

Fig. 4.1 Normalized interlaminar shear strength for specimens exposed at 40°C and stressed

and the error bars are the standard deviation of the average. In this case, all three humidity conditions show a slight increase in properties at the 2 week time interval. This is followed by a continuous degradation for the medium (50%) and wet (93%) humidity conditions at 4 and 6 weeks. The dry (2% RH) condition shows no discernable degradation over 4 and 6 week time intervals.

The data for the 85°C stressed cases at relative humidities of 2%, 50%, and 93% are plotted in Fig. 4.2. The data plotted is the average value of a series of 5 tests, and the error bars are the standard deviation of the average. In this case, the dry condition shows a continuous increase in interlaminar shear strength over the time intervals. The medium humidity condition shows a slight increase after 2 weeks,

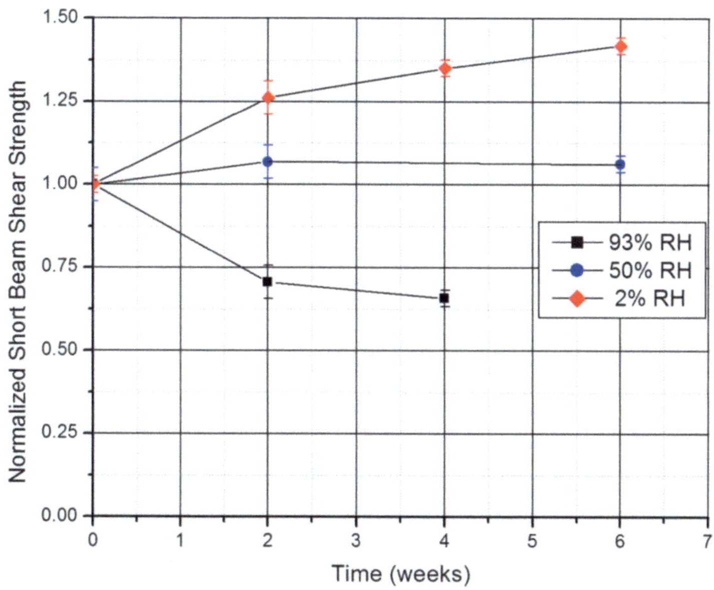

Fig. 4.2 Normalized interlaminar shear strength for specimens exposed at 85°C and stressed

Table 4.2 Short beam shear strength for specimens exposed with humidity control and unloaded

	100°C		125°C		150°C	
Baseline value (MPa)	46.4	Time (days)	46.4	Time (days)	46.4	Time (days)
Percent change	5.40	6	16.0	8	27.7	6
Percent change	12.1	12	12.7	18	27.3	14
Percent change	7.50	20	5.60	28	18.1	22

followed by a possible small decrease at 6 weeks. The wet condition shows a continuous decrease in properties over 2 and 4 weeks. Note that data was missing for the medium condition at 4 weeks and the wet condition at 6 weeks.

Based on these plots, the data seem to show a couple of mechanisms changing the properties during environmental exposure. First, the presence of moisture tends to decrease or degrade the properties with time. Second, there is an initial increase in properties, probably due to post curing of the epoxy matrix due to the elevated temperatures.

Additional short beam shear degradation testing was conducted to investigate and accelerate the thermal degradation mechanism past post cure. The temperatures chosen were 100°C, 125°C, and 150°C. These exposures were also conducted in the Cincinnati Sub Zero temperature and humidity cabinet; however, at these elevated temperatures the humidity control was turned off. With temperatures at and above 100°C, it is assumed that moisture is driven out of the chamber representing a dry exposure condition.

The degradation data for these high temperature exposure conditions are given in Table 4.2. This data is presented in the form of percent change of short beam shear

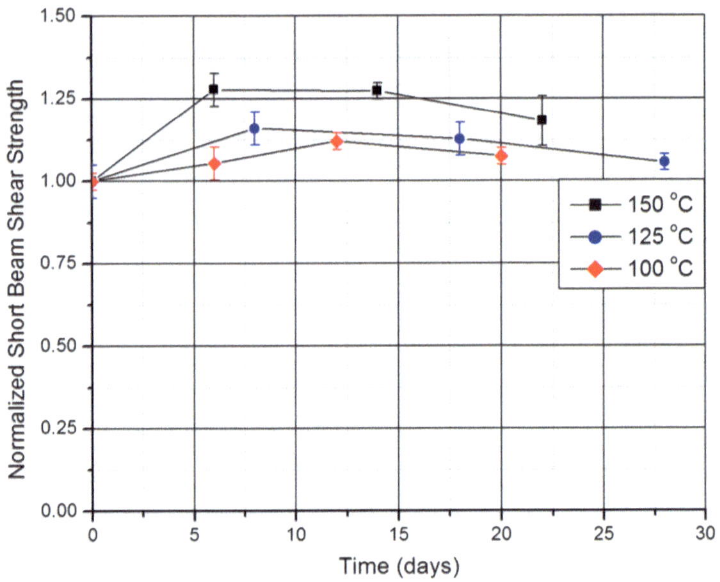

Fig. 4.3 Normalized interlaminar shear strength for specimens exposed at high temperatures without humidity control (dry)

strength compared to the baseline value as a function of time. This data for the cases at temperatures of 100°C, 125°C, and 150°C are plotted in Fig. 4.3. The data plotted is the average value of a series of five tests, and the error bars are the standard deviation of the average. For all three temperature conditions, there is an increase in interlaminar shear strength at the first time interval followed by a steady decrease in properties at subsequent time intervals. This indicates an additional temperature dependent degradation mechanism occurring.

Based on the data from the hygrothermal exposure tests, the modeling efforts will consider three degradation mechanisms: (1) "Post-curing," a thermally-driven reaction that causes increases in matrix properties, (2) thermal degradation, a separate, thermally-driven reaction causing destructive chemical change to the matrix, and (3) hygrothermal degradation, a third reaction dependent on temperature and moisture level, that is also destructive to the matrix material.

4.4 Modeling

Modeling the long term durability of a FRP composite material is a challenging task. FRP composites are made from high strength and modulus fibers bonded within a polymer matrix. The fibers come in many types, such as glass, carbon and aramid, and can be incorporated as unidirectional fibers, fabrics and mats. The polymer systems, typically thermosetting polymers, can be epoxies, polyesters, vinylesters and other formulations. There are several manufacturing techniques depending on the materials and the size and application of the components. Because of these material and processing variables, the degradation mechanisms are also complex and often coupled. As a result, the majority of FRP composite degradation studies have focused on a single mechanism, condition, or application.

The goal of this research is to develop a general degradation model, applicable to a wide range of FRP composites, to predict their service life based on environmental conditions. This type of prediction is needed because it is impractical to conduct long term aging tests that span the total service life of a structure. We want to be able to predict the long term degradation based on a minimum series of shorter term accelerated tests. Wherever possible, these tests would be conducted according to standardized test methods. In order to meet these goals, a chemical type model was used. In this approach, the degradation mechanisms are assumed to progress according to rate equations. For this model, we used an Arrhenius-based rate equation that incorporates the well known principle of time-temperature superposition.

The modeling approach and implementation was conceived by Dr. Hugh McManus of Metis Design Incorporated. A detailed account of this model development is provided in a companion paper by McManus et al. (2009), a summary of which is contained here. For the model, a semi-empirical approach is selected. The three reactions are tracked through degradation parameters a, b, and c, respectively. These parameters are dimensionless values from 0 (virgin state) to 1 (reaction/degradation fully completed). They are assumed to obey Arrhenius-type reaction laws:

$$\frac{\partial a}{\partial t} = k_a \left(1-a\right)^{n_a} \exp\left(\frac{-E_a}{RT}\right) \tag{4.1}$$

$$\frac{\partial b}{\partial t} = k_b \left(1-b\right)^{n_b} \exp\left(\frac{-E_b}{RT}\right) \tag{4.2}$$

$$\frac{\partial c}{\partial t} = k_c C_m^{n_m} \left(1-c\right)^{n_c} \exp\left(\frac{-E_c}{RT}\right) \tag{4.3}$$

where the k, n, and E parameters are empirical constants determined from tests.

The diffusion of moisture into the composites is assumed to be following Fickian's law from Shen and Springer (1976). The 3-D relation is:

$$\frac{\partial C_m}{\partial t} = \frac{\partial}{\partial x_i}\left(D_{ij}^m \frac{\partial C_m}{\partial x_i}\right) \tag{4.4}$$

where C_m is moisture concentration and D_{ij} is the diffusivity tensor. The boundary condition is:

$$C_\infty = A\left(\Phi\right)^b \tag{4.5}$$

where Φ is the environmental concentration (usually expressed as a relative humidity) and A and b are constants. The constant b is normally assumed to be 1, and A is then the percent moisture in the material at equilibrium with a saturated ($\Phi = 1$) environment, also referred to as the saturation moisture concentration.

The degradation is applied at the composite lamina level using micromechanics. In this case, the micromechanics of degradation approach developed by McManus and Chamis (1996) is used to calculate the properties of composite plies as the matrix material degrades under the reactions tracked in Eqs. 4.1–4.3 above. The progress of the three degradation mechanisms, as well as the moisture content (measured in dimensionless, 0-to-1 fashion by using the parameter C_m/A), are assumed to correspond to changes in room temperature material properties according the relations:

$$P_{RT} = P_{RTo}\left[1 + \Delta P_a a\right] \times \left[1 + \Delta P_b b\right] \times \left[1 + \Delta P_c c\right] \times \left[1 + \Delta P_m \left(\frac{C_m}{A}\right)\right] \tag{4.6}$$

The factors ΔP_i represent the percentage change in the property caused by completion of the corresponding reaction (or the absorption of moisture to saturation).

The properties of the composite in both the initial and the degraded state are calculated using Classical Laminated Plate Theory (CLPT). The degraded strength is calculated using progressive failure criteria as outlined by Tsai (1988).

The degradation models were implemented as a composite durability prediction tool through a software interface. This tool allows the user to define their laminate and service conditions, input material property data from a series of accelerated tests, and define a failure criterion for the laminate. The software tool then makes a prediction of how long it will take until the composite has degraded to the failure criterion.

An experimental test matrix was developed to determine the model parameters. This test matrix includes both short beam shear and tensile specimens. The tensile tests were conducted on both unidirectional and ±45° specimens. These tests were being conducted using an E-glass/Epoxy composite. To fully populate the coefficients in the degradation reaction equations, the specimens were exposed to five different exposure conditions. These conditions were: room temperature-dry, room temperature-wet, warm-wet, hot-dry and hot-wet. The temperatures are defined as follows: (1) room temperature is 23°C, (2) warm temperature is 43°C, and (3) hot temperature is 66°C. The room temperature dry samples were stored in sealed bags with desiccant at ambient temperature. The hot dry specimens were conditioned in a chamber assumed to be dry. All of the wet specimens were conditioned fully immersed in water. Specimens are being pulled and tested at four time intervals: 1 month, 3 months, 6 months, and 9 months. This data was entered into the software tool as input to the durability predictions.

4.5 Model Predictions

In order to test the predictions of the degradation model, two additional series of tests were conducted independently from the test matrix to determine model parameters. In this case, short beam shear and tensile tests were also conducted. The specimens were made from the same E-glass fabric and resin, but were

Table 4.3 Change in normalized mechanical properties for unidirectional and quasi-isotropic laminates

Exposure time (days)		Uniaxial tensile strength	Quasi-isotropic tensile strength	Interlaminar shear strength
Laminates exposed at 85°C and 50% relative humidity				
0	Baseline value (Mpa)	1.00	1.00	1.00
14	Percent change	0.50	0.58	0.96
28	Percent change	0.52	0.59	0.73
42	Percent change	0.53	0.56	0.70
Laminates exposed at 125°C and dry				
0	Baseline value (Mpa)	1.00	1.00	1.00
7	Percent change	0.98	1.03	1.17
14	Percent change	0.95	1.0	1.06
21	Percent change	0.98	0.98	1.21

manufactured, conditioned, and tested at a different site. In addition to unidirectional tension specimens, a quasi isotropic laminate was also included to test the capabilities of the software when applied to arbitrary laminates. The quasi isotropic layup was $[0, 90, \pm 45]_S$. The two exposure conditions were 85°C and 50% relative humidity, and 125°C and dry. For the 85°C and 50% relative humidity condition, samples were pulled and tested at 2, 4, and 6 weeks. For the 125°C and dry condition, samples were pulled at 1, 2, and 3 weeks.

The degradation data for this series of laminate tensile and interlaminar shear tests are given in Table 4.3. This data has been normalized against the baseline (unexposed) values to provide a comparison against the model predictions. The data for the 85°C and 50% relative humidity tensile tests are plotted in Fig. 4.4. The data plotted is the normalized average value of a series of five tests, and the error bars are the standard deviation of the average. For this case, the agreement with the model prediction is good. The model slightly under predicts the degradation in the tensile strength compared to the experimental results. In this case, the hygrothermal degradation mechanism dominates.

The data for the 85°C and 50% relative humidity interlaminar shear tests are plotted in Fig. 4.5. Again, the agreement with the model predictions is good. In this case, the model slightly over predicts the degradation in the inter-laminar shear strength compared to the experimental results. As with the tensile case, the hygrothermal degradation mechanism dominates for the interlaminar shear. However, this degradation mechanism is slower due to the fact that it takes more time for the moisture to diffuse into the thicker interlaminar shear specimens during exposure.

The data for the 125°C and dry tensile tests are plotted in Fig. 4.6. For this case, the model is predicting a slight increase in properties due to the post curing mechanism. The experimental data does not show any increase or decrease during exposure period. This is probably due to the progressive failure criterion predicting an increase in strength due to an increase in matrix properties that is not seen in the experimental values.

The data for the 125°C and dry interlaminar shear tests are plotted in Fig. 4.7. The agreement with the model predictions is good. In this case, the model predicts

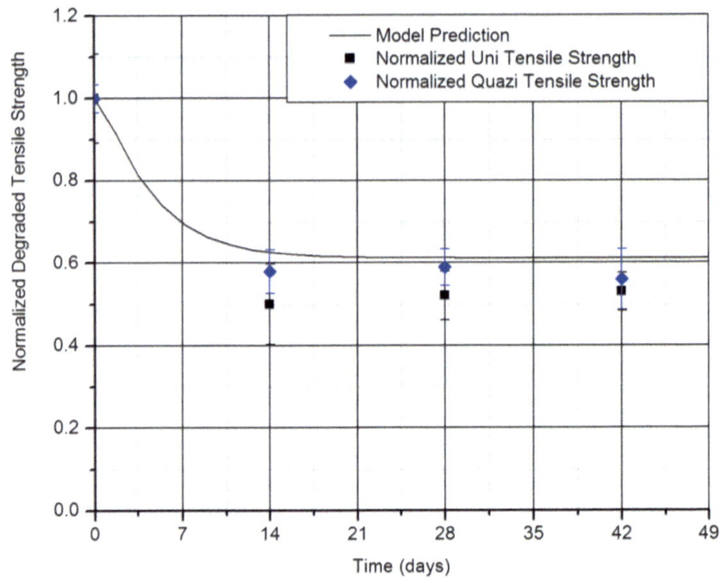

Fig. 4.4 Normalized tensile strength for unidirectional and quazi isotropic laminates exposed at 85°C and 50% relative humidity plotted against model predictions

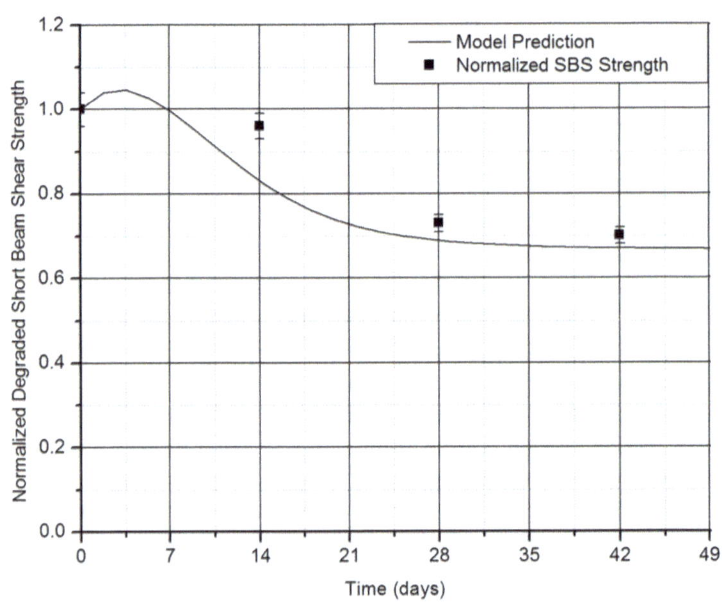

Fig. 4.5 Normalized interlaminar shear strength for unidirectional laminates exposed at 85°C and 50% relative humidity plotted against model predictions

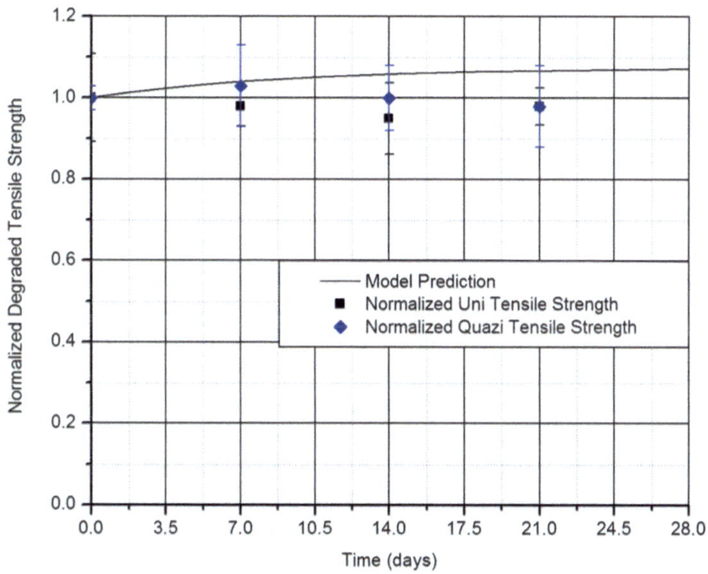

Fig. 4.6 Normalized tensile strength for unidirectional and quazi isotropic laminates exposed at 125°C and dry plotted against model predictions

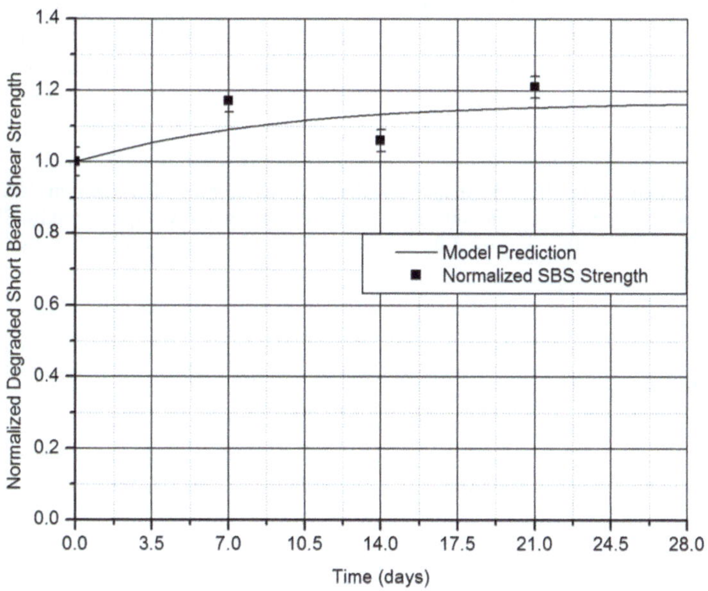

Fig. 4.7 Normalized interlaminar shear strength for unidirectional laminates exposed at 125°C and dry plotted against model predictions

an increase in interlaminar shear strength due to the post curing mechanism. While the experimental data does not track the model exactly, they do follow the general trend.

4.6 Conclusions

In this paper, we present results from accelerated hygrothermal degradation tests on E-glass/Epoxy FRP composite samples. From these accelerated tests, three primary degradation mechanisms were identified: (1) "Post-curing," a thermally-driven reaction that causes increases in matrix properties, (2) thermal degradation, a separate, thermally-driven reaction causing destructive chemical change to the matrix, and (3) hygro-thermal degradation, a third reaction dependent on temperature and moisture level, that is also destructive to the matrix material. A model was developed based on Arrhenius rate equations, and this modeling methodology was implemented through a software interface. Model predictions were compared against accelerated tests for arbitrary laminates and shown to predict the short term degradation relatively well.

While these results are promising and show the preliminary model worked well, additional work is needed to validate the usefulness of the model to predict long term behavior. First, longer term accelerated testing needs to be conducted to fully populate the reaction constants of the model. Additionally, long term exposure testing at lower temperatures needs to be conducted to determine the behavior of the FRP composite in actual service conditions. Finally, the model needs to be applied to other FRP composite systems consisting of different polymer matrices and reinforcement types.

Acknowledgments Dr. Hugh Mcmanus and Dr. Seth Kessler of Metis Design developed the modeling methodology and coding into a software tool. This work was sponsored by ERDC contract W9132T-09-C-0006 "Service Life Assessment Methodology for Composites (SLAM-C) Extended Study."

References

ASTM D 2344 (2006) Standard test method for short-beam strength of polymer matrix composite materials and their laminates. ASTM International, 100 Barr Harbor Drive, West Conshohocken

ASTM D 3039 (2010) Standard test method for tensile properties of polymer matrix composite materials. ASTM International, 100 Barr Harbor Drive, West Conshohocken

McManus HL, Chamis CC (1996) Stress and damage in polymer matrix composite materials due to material degradation at high temperatures. NASA Technical Memorandum 4682, Cleveland

McManus HL, Kessler S, Raghavan A, Hyer M, Case S, Cain J (2009) Service live assessment methodology for composites. Fiber Reinforced Polymer (FRP) composites for infrastructure applications conference, San Francisco, Nov 2009

Shen CH, Springer GS (1976) Moisture absorption and desorption of composite materials. J Compos Mater 10:2–20

Tsai SW (1988) Composite design, 4th edn, Think composites, Section 12

Chapter 5
Effects of Ultraviolet Radiation and Condensation on Static and Dynamic Compressive Behavior of Nanophased Glass/Epoxy Composites

Shaik Zainuddin, Mahesh Hosur, Ashok Kumar, and Shaik Jeelani

Abstract Increased use of fiber reinforced polymeric composites in an outdoor environment has led to questions concerning their environmental durability, particularly as related to ultraviolet (UV) radiation, moisture, and temperature exposure. This chapter describes the effects of UV and UV radiation + condensation (UC) on the static and dynamic compressive properties of unidirectional glass/epoxy composites. The samples were manufactured using an infusion process with and without nanophased epoxy and exposed to UV radiation and UC conditioning for 5, 10, and 15 days respectively. Nanophased epoxy was prepared with 1 wt%and 2 wt% nanoclay. Static compression tests were carried out using MTS test system under displacement control mode at a crosshead speed of 1.27 mm/min. Dynamic compression tests were carried out using modified Split Hopkinson Pressure Bar (SHPB) at different strain rates. The compressive strength and stiffness were evaluated as functions of strain rate. Results of the study showed that samples lost weight when exposed to UV radiation, whereas they gained weight when exposed to UC conditioning. Weight gain or loss was lower for nanophased composites when compared to neat samples. Static and high strain compressive properties reduced for all the nanophased samples when compared with room temperature samples. However, the loss in compressive properties was lowest in nanophased composites with 2 wt% nanoclay.

Keywords Glass/epoxy composites • Nanoclay • Ultraviolet radiation/condensation

---

S. Zainuddin (✉) • M. Hosur • S. Jeelani
Center for Advanced Materials, Tuskegee University,
1200 W. Montgomery Rd., Tuskegee Inst, Tuskegee, AL 36088, USA
e-mail: szainuddin@mytu.tuskegee.edu; hosur@tuskegee.edu; jeelanis@tusk.edu

A. Kumar
U.S. Army Engineer Research and Development Center, Construction Engineering Research Laboratory (ERDC-CERL), Champaign, IL 61821-9005, USA
e-mail: Ashok.kumar@usace.army.mil

R. Jain and L. Lee (eds.), *Fiber Reinforced Polymer (FRP) Composites for Infrastructure Applications*, Strategies for Sustainability, DOI 10.1007/978-94-007-2357-3_5,

5.1 Introduction

Fiber reinforced polymer (FRP) composites are known for their high specific strength and stiffness, corrosion resistance and superior fatigue performances compared to metallic materials and hence are ideal materials for consideration in many applications, including civil infrastructure as reported by Malvar (1996) and Nairn (1993). Significant weight and long-term cost savings can be realized by replacing steel with composites in these applications. However, the drawback of polymeric composites is their inherent viscoelastic behavior. Microstructural changes, time-dependent deformation, and degradation in mechanical properties are known to affect the performance in FRP composites when exposed to environmental conditions such as extreme temperatures and moisture (Bonniau and Bunsell 1981, Weitsman and Guo 2002). Polymer composites used in outdoor applications are exposed to ultraviolet radiation and are susceptible to photo-oxidation, leading to the surface degradation as noted by Giori and Yamauchi (1984) and Haliwell (1992). The typical wavelength of UV radiation is in the range of 290–400 nm. The energy associated with the UV radiation is similar to the dissociation energy of most of the polymers, which is in the range of 290–460 kJ/mole. Exposure to the UV radiation over time causes an increased scission and crosslinking of polymers causing their degradation over time. The durability of polymer composites in such an environment is one of the primary issues limiting their acceptance in many applications, especially in infrastructural applications.

Recent advancement in modifying the matrix properties, with the addition of a small weight percentage of nanoparticles, has shown significant improvement in the strength and the durability of both the matrix and FRP composites. Inorganic nanoparticles have gained acceptance as potential reinforcing materials because of their low cost and ease of fabrication (LeBaron et al. 1999 and Alexandre and Dubois, 2000). Many researchers attempted to enhance properties of different types of polymers by the addition of nanoclays. Addition of nanoclay increases the barrier properties and enhances interfacial bonding (Maul 2005). Woo et al. (2008) investigated environmental degradation of epoxy nanocomposites due to the exposure UV radiation. They reported wider and shallower cracks with less degree of discoloration in epoxy nanocomposites over neat counterparts. They credit this result to the excellent barrier characteristics of the organoclay with high aspect ratio. Similar results were also reported by Nanocor Inc. (Maul 2005). Researchers demonstrated the possibility of functionalizing the surface of layered inorganic silicates to make them compatible with the polymeric matrix as reported by Fukushima and Inagaki (1987) and Usuki et al. (1993). Hackman and Hollaway (2006) studied potential applications of clay nanocomposite materials to civil engineering structures. They concluded that their ability to increase service life of materials subjected to aggressive environments could be utilized to increase the durability of glass and carbon fiber composites.

Though there has been some work in improving the strength, stiffness and durability of laminated composites through the inclusion of nanoclay, not been much work has been done to determine their response to UV radiation with or without condensation. Hence, the objective of the work reported here is to improve the matrix by utilizing the advantages of nanoclay and to study the durability of FRP composites subjected to static and dynamic compressive loading following the exposure to ultraviolet radiation and condensation.

5.2 Materials and Fabrication

A two-phased toughened SC-15 epoxy resin, Nanomer® I. 28E nanoclay and unidirectional E-glass fibers were used in this study. SC-15 is a room temperature curing epoxy resin with a viscosity of 300 cps and a pot life of 6 h. Low viscosity and longer pot life make it more compatible to use as matrix in FRP composites. Nanomer® I-28E nanoclay is a naturally occurring montmorillonite—a 2–1 layered clay mineral with a plate like structure. This nanoclay has a thickness of the order of 1 nm and lateral dimensions of several microns, thereby providing a large aspect ratio. In order to interact chemically with the epoxy system, it is modified by Nanocor Inc. with proprietary functional groups.

5.2.1 Modification of Epoxy Resin Using Nanoclay

A measured amount of nanoclay was first preheated at 100°C for 2 h and degassed to avoid the lump formation due to the moisture absorption from ambient conditions. Nanoclay was then dispersed in the part A of SC-15 epoxy resin with a magnetic stirrer for 2 h. The modified part A was then mixed with the part B of SC-15 epoxy at a ratio of 10:3 using a high speed mechanical stirrer at about 800 rpm for 5 min. The intense mechanical mixing produces air bubbles in the mixture, which if not taken out will increase void content in the composite. These trapped air bubbles were taken out by keeping the mixture in a high vacuum dessicator chamber for about 30 min. Once the entrapped bubbles were removed, the mixture was poured into a beaker and left for infusion.

5.2.2 Fabrication of Glass Fiber Reinforced Polymer (GFRP) Composites

The vacuum assisted resin infusion molding (VARIM) process was used for fabricating unidirectional E- glass/epoxy composites. The VARIM process uses vacuum pressure to remove air from the fabric lay-up before and while the matrix resin is introduced to the fabric reinforcement. The pressure difference between the atmosphere and the vacuum is the driving force for infusion of the resin into the lay-up. The nanophased resin mixture was infused through the inlet section into 26 layers of unidirectional E-glass fabric and left for curing at room temperature for 24 h.

5.3 Environmental Conditioning

5.3.1 Ultraviolet Radiation (UV) and UV + Condensation (UC) Conditioning

Degradation mechanisms resulting from ultraviolet radiation (UV) and ultraviolet radiation + condensation (UC) were studied by subjecting the GFRP panels to UV and UC conditioning in a QUV/Se weathering chamber (Q-Panel Lab Products, Cleveland Ohio). Samples of nominal size 60 mm × 12.5 mm × 5 mm cut from GFRP panels were exposed to two different conditions:

1. Exposure to UV radiation only
2. Sequential exposure to UV radiation for 4 h followed by condensation for 4 h

Under each condition, panels were exposed to 120 h (5 days), 240 h (10 days), and 360 h (15 days). In addition, panels cured at room temperature (RT) were also used to generate baseline data.

5.4 Experimental Procedures

5.4.1 Weight Gain Analysis

The samples were weighed with a Mettler AT250 digital balance (precision 0.01 mg). The percent weight gain as a function of time was calculated using Eq. 5.1:

$$\%Weightgain = \frac{w_f - w_i}{w_f} X100 \tag{5.1}$$

Where w_f is the final weight of the sample and w_i is the initial weight of the dry sample at room temperature.

5.4.2 Static and Dynamic Compression Tests

In order to investigate the quasi-static and dynamic compression response, an average of ten specimens were tested in the fiber direction. ASTM D695-08 (2002) was followed for quasi-static compression test. The test was carried out in displacement control mode at a strain rate of 10^{-4}/s. Dynamic compression test was also performed at three different strain rates (382/s, 440/s, 510/s) using split Hopkinson pressure bar (SHPB). The dimensions of a sample for both the tests were 12.5 × 12.5 × 5 mm. In SHPB, the sample was sandwiched between two long bars, 150 cm in length and 38 mm in diameter, called incident and transmission bars, which were instrumented with strain gages mounted on the surface at their mid-length. During the test, a compressive pulse was imparted to the incident bar by propelling a striker bar of the same material and diameter with a length of 30 cm from a pressurized chamber. The compressive pulse traveled through the incident bar. Part of the energy associated with the pulse was reflected back into the incident bar due to the acoustic impedance mismatch between the bar material and the sample. The rest of the energy was transmitted into the transmission bar through the sample. Dynamic stress-strain relation was derived from incident, reflected and transmitted pulses assuming one-dimensional wave propagation theory. The equations for the strain rate ($\acute{\varepsilon}$), strain (ε), and stress (σ) of the specimen are given by Meyers (1994):

$$\dot{\varepsilon}(t) = -\frac{2C_0}{L_s}\varepsilon_R(t) \tag{5.2}$$

$$\varepsilon(t) = -\frac{2C_0}{L_s}\int_0^t \varepsilon_R(t)dt \tag{5.3}$$

$$\sigma(t) = \frac{\mathrm{E}_b\mathrm{A}_b}{2\mathrm{A}_s}[\varepsilon_I(t) + \varepsilon_R(t) + \varepsilon_T(t)] \text{ or } \sigma(t) = \frac{\mathrm{E}_b\mathrm{A}_b}{\mathrm{A}_s}\varepsilon_T(t) \tag{5.4}$$

Since

$$\varepsilon_I(t) + \varepsilon_R(t) = \varepsilon_T(t) \tag{5.5}$$

where $C_0 = \sqrt{\mathrm{E}_b / \rho_b}$, is the longitudinal wave velocity in the incident bar, L_s is the specimen length and $\varepsilon_R(t)$, $\varepsilon_I(t)$, and $\varepsilon_T(t)$ are the time-dependent reflected, incident and transmitted strains, E_b, A_b, and ρ_b, and E_s, A_s, and ρ_s are the modulus, cross-sectional area and density of the bar and specimen respectively. Different strain rates are obtained by adjusting the pressure in the chamber through which the striker bar is propelled.

5.5 Results and Discussion

5.5.1 Weight Gain/Loss Due to UV and UC Exposure

Figure 5.1 shows the weight gain/loss (%) vs. number of days of neat, 1 wt% and 2 wt% specimens exposed to UV and UC conditions. Samples exposed to UV conditions showed loss in weight with respect to time with a maximum of 0.17% in neat, 0.16% in 1 wt% and 0.12% in 2 wt% samples after 15 days conditioning as shown in Fig. 5.1. However, after approximately 12 days (216 h), loss in weight reached almost the saturation point and showed minimal decrease with respect to time. In polymers, chain scission will produce small molecules when exposed to UV radiation. These small molecules capped by oxygen from air are free to migrate out from the specimen, causing weight loss. The polymer resin used here is diglycidylether

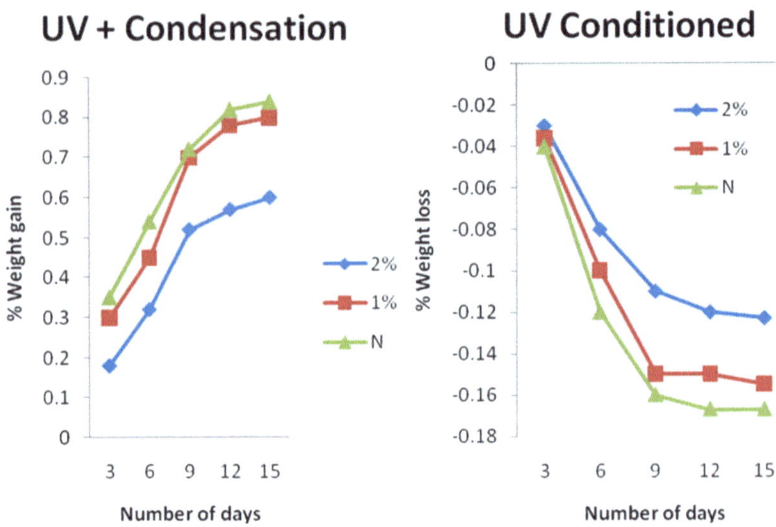

Fig. 5.1 Change in specimen weight as a function of time exposed to UV radiation

type that has less radiation resistance, which might be due to the lower level crosslinking density for the diglycidyl type of epoxy resin as reported by Liau et al. (1998). Therefore, oxygen can easily diffuse into the polymer from air and accelerate the degradation of the resin.

Specimens exposed to alternate ultraviolet radiation and condensation showed gain in weight with respect to time with a maximum of 1.15% in neat, 1.06% in 1 wt% and 0.88% in 2 wt% samples after 15 days conditioning shown in Fig. 5.2. However, water absorption almost reached saturation level in 1 and 2 wt% UC condition samples and showed no further gain with respect to time after approximately 9 days of conditioning. However, weight of neat samples was found to increase even after 15 days of conditioning. The increase in weight in UC condition samples can be attributed to a combined effect of ultraviolet radiation and moisture. Ultraviolet irradiation produces microcracks on the surface and provides pathway for rapid ingress of moisture. Also, the presence of moisture in form of water molecules enhances photo oxidation reactions due to increased concentration of OH^- and H^+ ions by Kumar et al. (2002). These water molecules may move on to the free volume sites between polymer molecules, improving the mobility of free radicals within the surface layer, thus accelerating photo-oxidation by Tjandraatmadja et al. (2002). Condensation due to the presence of water can also remove soluble products of photo oxidation reactions from an ultraviolet irradiated surface and thereby expose fresh surfaces that will be susceptible for further degradation by ultraviolet radiation. However, 2 wt% samples were seen to be least affected by both UV and UC conditioning. The loss in weight in UV and weight gain in UC conditioned samples were lower in case of 2 wt% samples in comparison to neat and 1 wt% loaded samples. This can be attributed to the gas/moisture and radiation barrier properties of montmorillonite nanoclay, which is consistent with results reported by

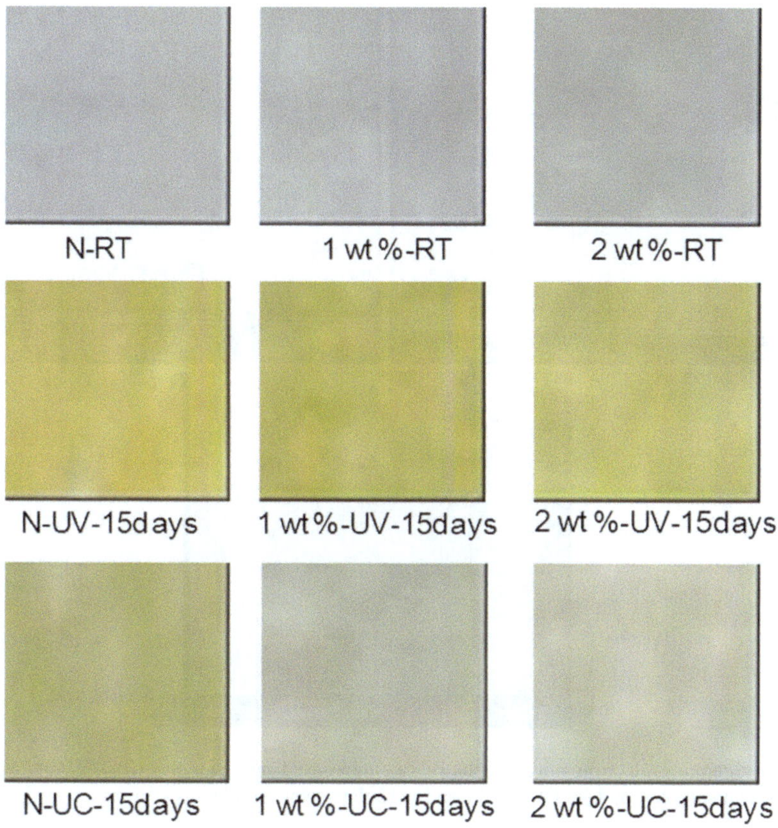

Fig. 5.2 Color change due to UV and UC conditioning

Ricky et al. (2007). Also, the addition of nanoclay can cause possible catalytic action that accelerates curing (Carrasco and Pages, 2008) and increases crosslinking between polymer chains, resulting in higher cross linking density (Uhl et al. 2004 and Yuanxin et al. 2006). Due to this enhancement in crosslinking of polymer chains, damaging effects of environmental conditions may be subsided and thereby improve the overall performance as reported by Ricky et al. (2007) and Zainuddin et al. (2009).

5.5.2 *Surface Morphology*

The surfaces of all the specimens exposed to UV radiation and UC conditions exhibited distinct change in color from opaque white to yellowish white during early stages of the exposure. Since UV light cannot penetrate deeply into the samples, discoloration was observed only on the exposed surface of the sample. This discoloration might be due to the photo-oxidation that resulted in the formation of chromic chemical species, which absorbed the visible range of light (Kumar et al. 2002).

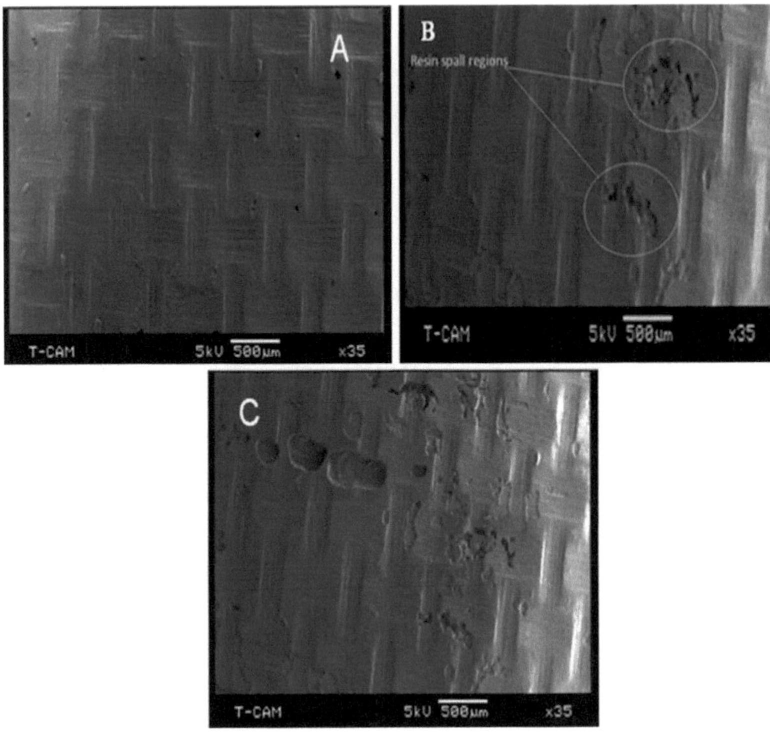

Fig. 5.3 Scanning electron micrographs of (**a**) unexposed neat sample, (**b**) UV condition for 15 days neat samples, and (**c**) UC condition for 15 days neat samples

Minor changes in surface roughness were also visible for all the specimens. Exposure to UV + condensation resulted in less discoloration in samples in comparison to UV conditioned samples and no visible changes were observed in surface morphology. The extent of color change was considerably reduced in 2 wt% samples in comparison to neat samples, as shown in Fig. 5.2. This can be attributed to barrier properties of montmorillonite clay reducing the moisture and radiation absorption. Further details regarding the physical processes that govern polymer degradation were revealed from the examination of specimens under scanning electron microscopy. Figure 5.3a–c shows the scanning electron micrographs of neat samples subjected to RT, UV, and UC conditioning for 15 days respectively, clearly indicating resin spalling from the sample surface. Higher magnification micrographs of top surface (exposed surface) of neat and 2 wt% samples subjected to UV and UC conditioning are shown in Figs. 5.4 and 5.5a, b. Samples subjected to UV conditioning revealed matrix microcracking, whereas UC conditioned samples revealed matrix microcraking and micropores. Microcraking was caused by the polymer matrix becoming more brittle due to the increased crosslinking resulting from photo-oxidation reaction induced by UV radiation, whereas micropores may have developed due to condensation effect. Similar result of matrix microcracking and void formation was reported by Kumar et al.

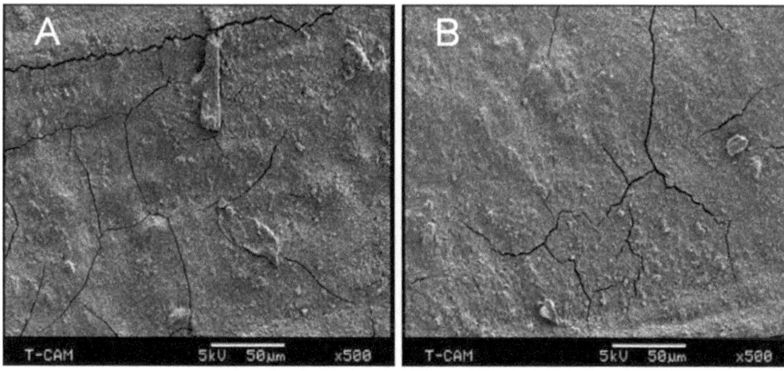

Fig. 5.4 SEM images of (**a**) neat sample subjected to UV conditioning for 15 days, (**b**) 2 wt% sample subjected to UV conditioning for 15 days

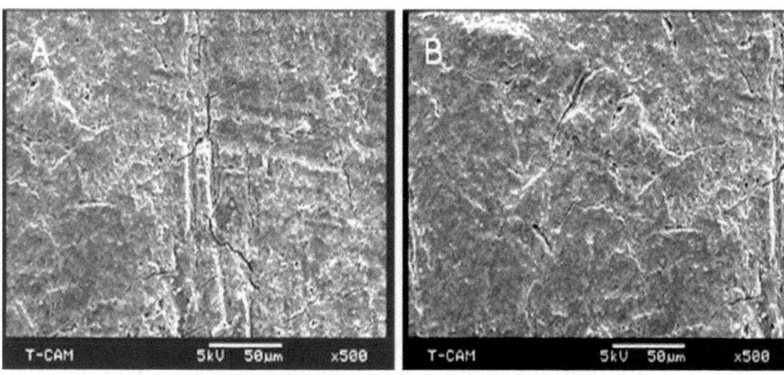

Fig. 5.5 SEM images of (**a**) neat sample subjected to UC conditioning for 15 days, (**b**) 2 wt% sample subjected to UC conditioning for 15 days

(2002). However, this degradation in both of the conditioned samples was found only on the top portion and no differences were observed on the remaining portion indicating that the effect of UV and UC conditioning is limited at the most subsurface level.

5.5.3 Static and Dynamic Compressive Properties: Stress-Strain Response

5.5.3.1 Room Temperature and Ultraviolet Radiation Conditioned Samples

Typical stress-strain curve and the data obtained from experiments for RT (baseline) and UV conditioned specimens are shown in Fig. 5.6 and Table 5.1, respectively. Compressive strength and modulus of 2 wt% samples were found to be higher in

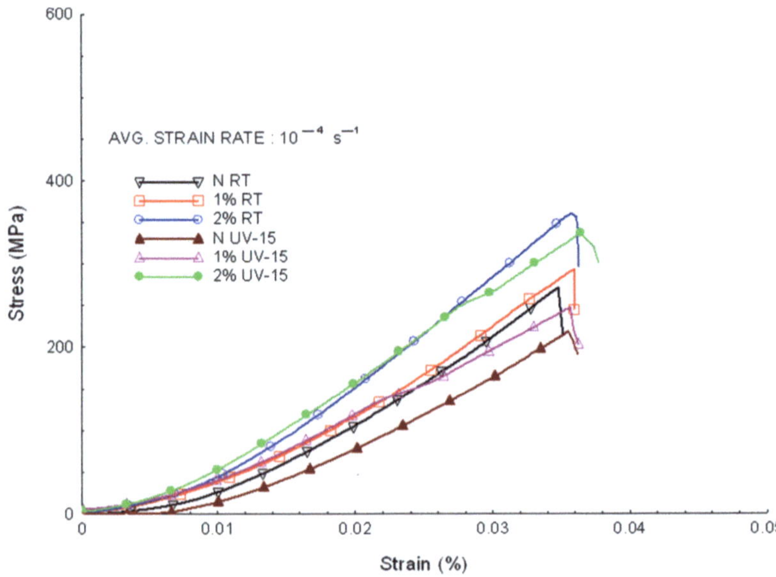

Fig. 5.6 Stress- strain curve of samples tested at baseline (room temperature) and after 15 days of UV conditioning at a strain rate of 10^{-4}/s

comparison to neat and 1 wt% samples. An increase of 32% and 24.5% in strength and modulus was observed in 2 wt% samples in comparison to neat samples, both conditioned at room temperature and tested at a strain rate of 10^{-4}/s. Similarly, an increase of 58.5% and 37% in strength and modulus was observed in 2 wt% samples over neat counterpart, both subjected to UV conditioning for 15 days and tested at a strain rate of 10^{-4}/s. However, 1 wt% samples showed nominal increase in these properties in comparison to neat samples similarly conditioned. Fractured samples were analyzed to understand the reason for improvements in 2 wt% samples in comparison to neat and 1 wt% samples similarly conditioned. Figure 5.7a–f shows the intensity of failure and fracture behavior of samples subjected to UV conditioning for 15 days and tested at a strain rate of 10^{-4}/s. Upon reaching the maximum peak stress, catastrophic failure was observed. Neat samples failed due to crushing, whereas the 1 wt% failed due to kink band formations which lead to crack propagation and fracture of the sample. However, 2 wt% samples failed due to fracture on one side and no delamination was observed. Similar failure behavior was observed in all the room temperature samples.

Modulus and strength of all the conditioned samples were found to increase with increasing strain rate with a maximum increase at a strain rate of 510/s as shown in Fig. 5.8 and Table 5.1. An increase of 30% and 21% in strength and modulus was observed in 2 wt% samples in comparison to neat samples both conditioned at room temperature. Additionally, an increase of 31.7% and 43% in strength and modulus was observed in 2 wt% UV condition- 15 days samples in comparison to neat counterpart similarly conditioned. Fracture behavior of samples

Table 5.1 Compressive properties of baseline (RT) and UV conditioned samples at different strain rates

		Static		Dynamic					
Condition	Sample	10^{-4} s^{-1}	% change	Avg. 382 s^{-1}	% change	Avg. 440 s^{-1}	% change	Avg. 510 s^{-1}	% change
Compressive strength (MPa)									
	Neat	273 ± 8	–	338 ± 12	–	383 ± 12	–	416 ± 18	–
RT	1%	293 ± 6.7	–	375 ± 15	–	436 ± 15	–	487 ± 24	–
	2%	361 ± 4.2	–	424 ± 11	–	466 ± 21	–	541 ± 22	–
	Neat	270 ± 5.5	−1.1	333 ± 16	−1.5	377 ± 11	−1.6	413 ± 17	−0.7
UV-5	1%	286 ± 3.8	−2.4	372 ± 21	−0.8	430 ± 18	−1.4	481 ± 12	−1.2
	2%	359 ± 7.6	−0.6	422 ± 23	−0.5	465 ± 12	−0.2	534 ± 15	−1.3
	Neat	255 ± 4.7	−6.6	313 ± 17	−7.4	335 ± 25	−12.5	396 ± 25	−4.8
UV-10	1%	264 ± 3.1	−9.9	355 ± 08	−5.3	408 ± 22	−6.4	458 ± 12	−5.9
	2%	353 ± 5.4	−2.2	422 ± 12	−0.5	458 ± 19	−1.7	504 ± 18	−6.8
	Neat	212 ± 5.6	−22	308 ± 17	−8.9	332 ± 12	−13	378 ± 17	−9
UV-15	1%	240 ± 7.3	−18	355 ± 5	−5.3	406 ± 17	−6.9	446 ± 21	−8.4
	2%	336 ± 5.5	−6.9	417 ± 23	−1.7	455 ± 11	−2.4	498 ± 22	−7.9
Compressive modulus (GPa)									
	Neat	10.2 ± 0.6	–	18.5 ± 1.2	–	18.7 ± 1.5	–	21.5 ± 2.6	–
RT	1%	10.2 ± 0.5	–	20.6 ± 0.7	–	21.5 ± 1.1	–	25.6 ± 1.2	–
	2%	12.7 ± 1.6	–	23.5 ± 1.1	–	24.5 ± 0.6	–	26 ± 2.3	–
	Neat	9.3 ± 0.6	−8.8	18.5 ± 1.3	0	18.7 ± 0.7	0	19.2 ± 1.8	−10.7
UV-5	1%	9.34 ± 0.8	−8.4	19.2 ± 0.5	−6.8	21.3 ± 1.2	−0.9	24.5 ± 2.1	−4.3
	2%	12.7 ± 1.6	0	22.5 ± 0.8	−4.2	24.2 ± 1.5	−1.2	25.6 ± 1.1	−1.5
	Neat	8.6 ± 0.5	−15.7	17.5 ± 1.2	−5.4	18.2 ± 1.2	−2.7	19 ± 2.4	−11.6
UV-10	1%	8.9 ± 0.7	−12.7	18.4 ± 0.7	−11	20.5 ± 1.8	−4.7	22.5 ± 1.5	−12
	2%	12.5 ± 0.4	−1.57	21 ± 0.4	−11	23.2 ± 2.1	−5.3	25.6 ± 1.8	−1.5
	Neat	8.54 ± 0.3	−16.3	16.3 ± 0.8	−12	17 ± 1.7	−9.1	17.5 ± 1.7	−18.6
UV-15	1%	8.6 ± 0.3	−15.7	17.9 ± 0.6	−13	19.5 ± 0.9	−9.3	20.2 ± 1.2	−21
	2%	11.7 ± 0.5	−7.9	21 ± 1.1	−11	23 ± 1.1	−6	25 ± 2.1	−3.6

% change: percent change in comparison to RT samples

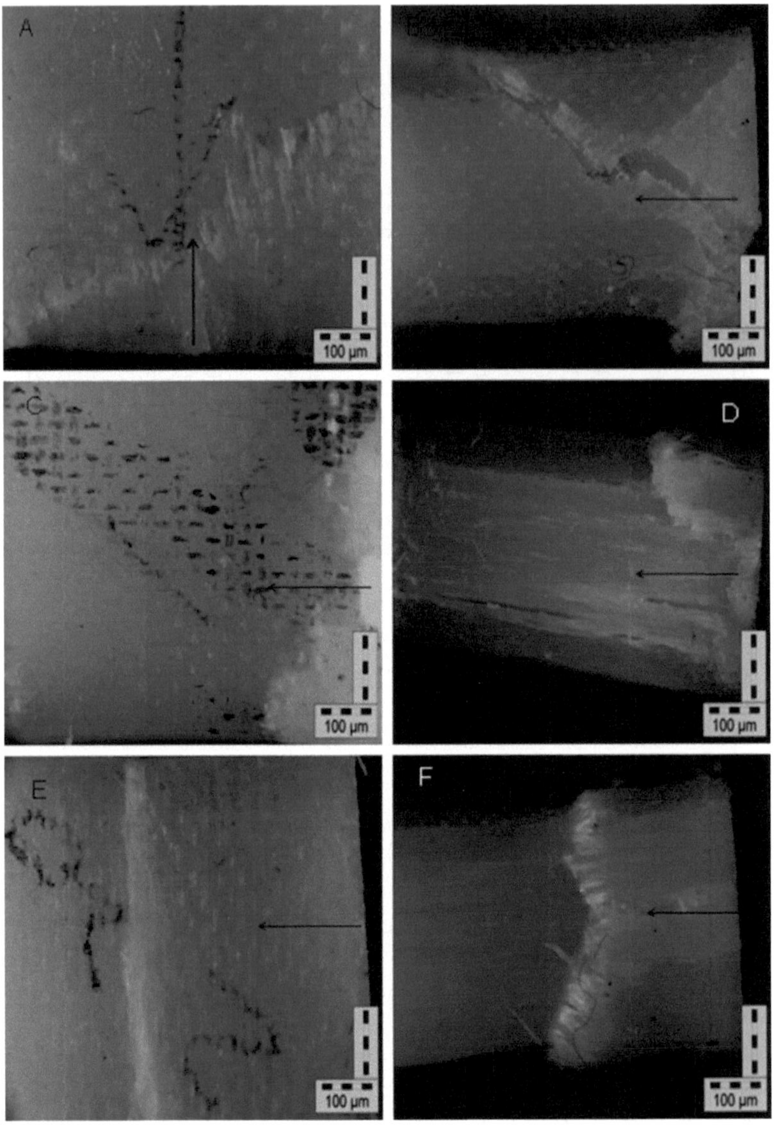

Fig. 5.7 Fracture behavior of samples subjected to UV conditioning for 15 days and tested at a strain rate of 10^{-4}/s, (**a**, **b**) Neat, (**c**, **d**) 1 wt% and, (**e**, **f**) 2 wt%

subjected to UV conditioning for 15 days and tested at a strain rate of 382/s is shown in Fig. 5.9, which shows that neat and 1 wt% samples subjected to UV conditioning for 15 days were partially crushed on the loading side and delaminated. 2 wt.% samples subjected to 15 days UV conditioning showed strong interfacial bonding and minimal crushing.

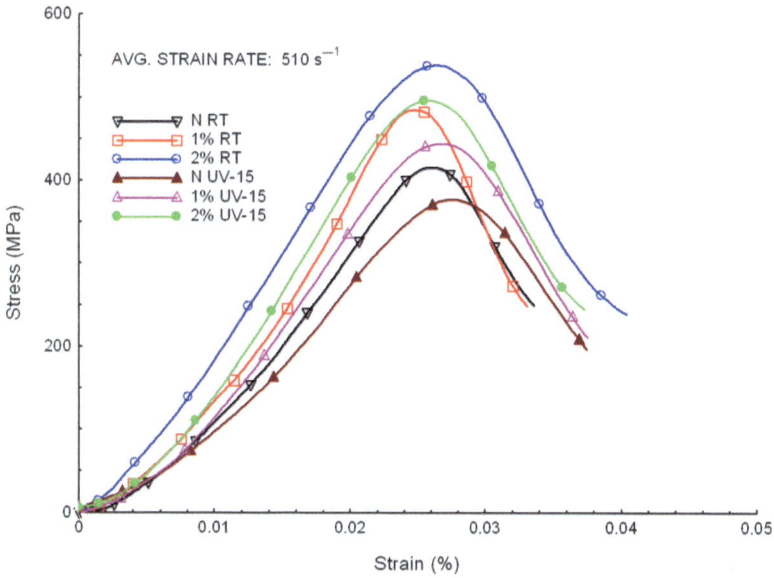

Fig. 5.8 Stress-strain curve of samples tested at baseline (room temperature) and after 15 days of UV conditioning at a strain rate of 510/s

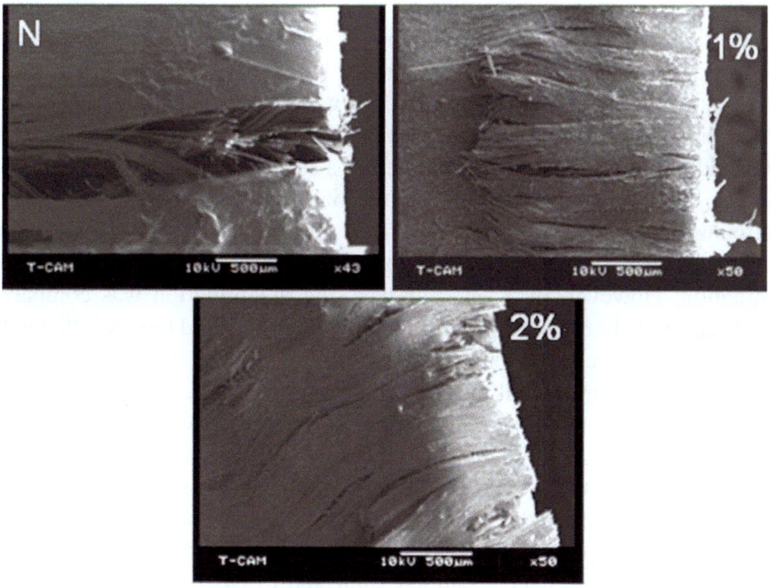

Fig. 5.9 Fracture behavior of neat, 1 wt% and 2 wt% samples subjected to UV conditioning for 15 days and tested at a strain rate of 382/s

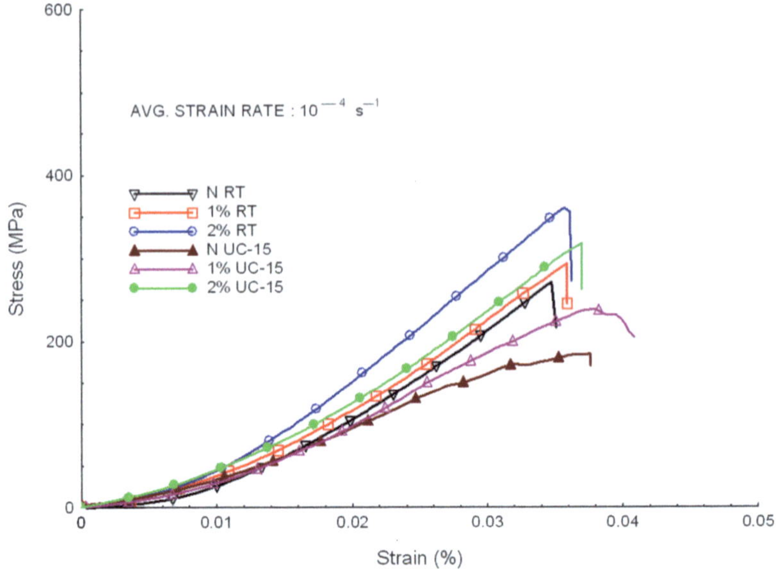

Fig. 5.10 Stress-strain curve of samples tested at baseline (room temperature) and after 15 days of UC conditioning at a strain rate of 10^{-4}/s

5.5.3.2 Alternate Ultraviolet Radiation and Condensation (UC) Conditioned Samples

Neat and nanoclay infused samples exposed to alternate ultraviolet radiation and condensation conditions also showed similar behavior as in only UV conditioned samples shown in Fig. 5.10 and Table 5.2. An increase of 67% and 37.3% in strength and modulus was observed in 2 wt% samples over neat counterpart, both subjected to UC conditioning for 15 days and tested at a strain rate of 10^{-4}/s. Figure 5.11a–f shows the intensity of failure and fracture behavior of samples subjected to UC conditioning for 15 days and tested at 10^{-4}/s strain rate. Neat and 1 wt% samples failed due to initial crushing on the one side followed by significant fiber/matrix delamination. However, in 2 wt% samples, the delamination was comparatively less and occurred after the samples reached the peak load and thus resulted in failure mainly due to fracture on one side of sample.

Modulus and strength of all the conditioned samples were found to increase with increasing strain rate with a maximum increase at a strain rate of 510/s, as shown in Fig. 5.12 and Table 5.2. An increase of 31% and 47% in strength and modulus was observed in 2 wt% UC 15 days conditioned samples in comparison to neat counterpart similarly conditioned. Figure 5.13 shows that all the samples subjected to UC conditioning for 15 days failed due to crushing followed by severe fiber/matrix delamination. However, 2 wt% samples subjected to UC conditioning for 15 days showed better interfacial bonding in comparison to neat and 1 wt% samples similarly conditioned as shown in Fig. 5.14.

Table 5.2 Compressive properties of baseline (RT) and UC conditioned samples at different strain rates

		Static		Dynamic					
Condition	Sample	10^{-4} s^{-1}	% change	Avg. 382 s^{-1}	% change	Avg. 440 s^{-1}	% change	Avg. 510 s^{-1}	% change
Compressive strength (MPa)									
	Neat	273 ± 8	–	338 ± 12	–	383 ± 12	–	416 ± 18	–
RT	1%	293 ± 6.7	–	375 ± 15	–	436 ± 15	–	487 ± 24	–
	2%	361 ± 4.2	–	424 ± 11	–	466 ± 21	–	541 ± 22	–
	Neat	260 ± 5.5	−4.8	323 ± 17	−5	367 ± 14	−4.2	390 ± 08	−6.3
UC-5	1%	276 ± 6.8	−5.8	352 ± 08	−6	414 ± 23	−5	473 ± 14	−2.9
	2%	346 ± 8.6	−4	422 ± 13	−0.5	445 ± 17	−4.5	522 ± 07	−3.5
	Neat	203 ± 9	−25.6	305 ± 18	−9.8	340 ± 16	−11	379 ± 24	−8.9
UC-10	1%	255 ± 5.1	−13	344 ± 23	−8.3	396 ± 25	−9	465 ± 21	−4.5
	2%	343 ± 8.4	−5	393 ± 26	−7.3	433 ± 12	−7	499 ± 18	−7.8
	Neat	190 ± 9.6	−30.4	292 ± 13	−13.6	320 ± 16	−16.5	377 ± 12	−9.4
UC-15	1%	239 ± 4.3	−18	330 ± 24	−12	384 ± 19	−11.9	451 ± 23	−7.4
	2%	319 ± 6.5	−12	382 ± 15	−10	425 ± 21	−8.8	494 ± 22	−8.7
Compressive modulus (GPa)									
	Neat	10.2 ± 0.6	–	18.5 ± 1.2	–	18.7 ± 1.5	–	21.5 ± 2.6	–
RT	1%	10.2 ± 0.5	–	20.6 ± 0.7	–	21.5 ± 1.1	–	25.6 ± 1.2	–
	2%	12.7 ± 1.6	–	23.5 ± 1.1	–	24.5 ± 0.6	–	26 ± 2.3	–
	Neat	9.2 ± 0.6	−9.8	17.7 ± 0.5	−4.5	18.2 ± 1.9	−2.67	20.2 ± 1.4	−6
UC-5	1%	9.2 ± 0.8	−9.8	19.5 ± 1.3	−5.3	20.3 ± 2.6	−5.58	24.5 ± 2.3	−4.3
	2%	12.7 ± 1.6	0	22 ± 0.6	−6.4	23.4 ± 2.4	−4.49	25.3 ± 1.8	−2.7
	Neat	7.8 ± 0.5	−23.5	16.9 ± 1.7	−8.6	17.4 ± 2.2	−7	19 ± 1.4	−11.6
UC-10	1%	9 ± 0.7	−11.8	19.2 ± 2.1	−6.8	19.5 ± 1.5	−9.3	23.5 ± 0.6	−8.2
	2%	11.7 ± 0.4	−7.9	22 ± 0.5	−6.38	22.4 ± 0.6	−8.6	25.2 ± 2.1	−3.1
	Neat	7.5 ± 0.3	−26.5	16.3 ± 1.4	−12	16.5 ± 1.4	−11.8	17 ± 1.3	−21
UC-15	1%	8.6 ± 0.3	−12.8	18.8 ± 0.4	−8.7	19 ± 2.1	−11.6	20 ± 1.6	−21.8
	2%	11.2 ± 0.5	−18.9	21.8 ± 1.7	−7.2	22 ± 2.5	−10.2	25 ± 2.1	−3.85

% change: percent change in comparison to RT samples

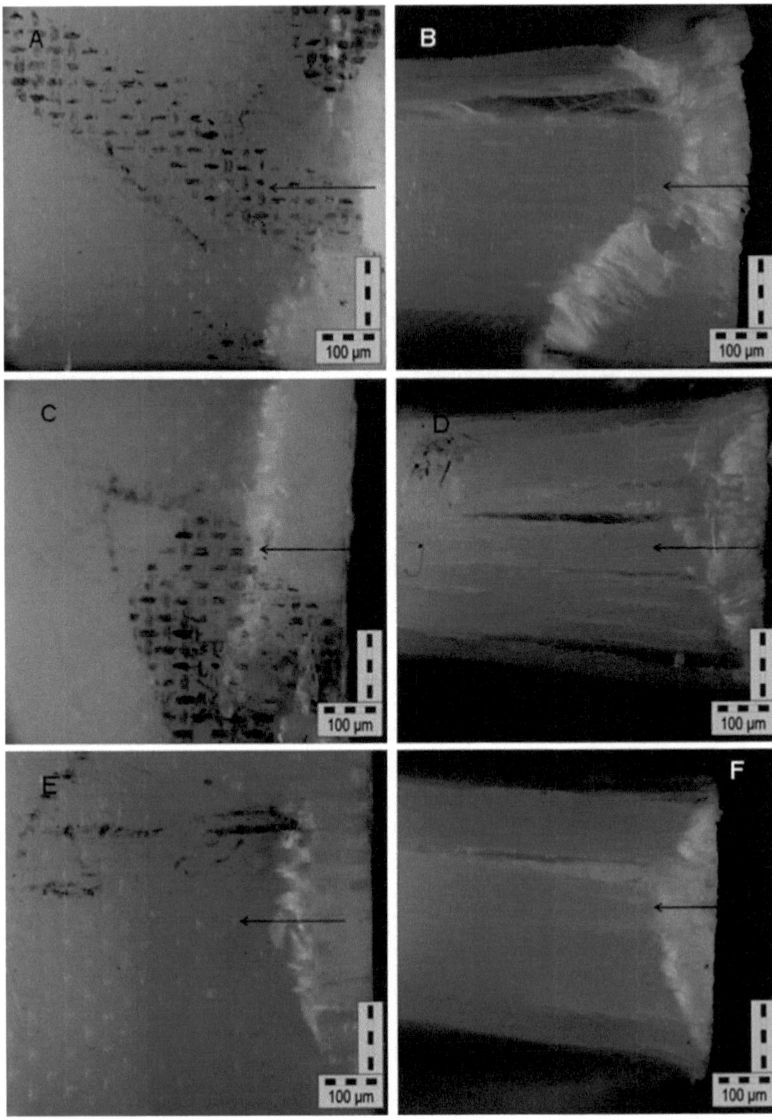

Fig. 5.11 Fracture behavior of samples subjected to UC conditioning for 15 days and tested at a strain rate of 10^{-4}/s, (**a**, **b**) Neat, (**c**, **d**) 1 wt% and, (**e**, **f**) 2 wt%

The extent of damage was significantly reduced in 2 wt% samples subjected to UV and UC conditioning for 15 days in comparison to neat and 1 wt% samples similarly conditioned. Lesser damage area and better interfacial bonding in nanoclay infused samples was evident from the weight gain study, scanning electron and optical micrographs directly relate to the enhancement in mechanical properties.

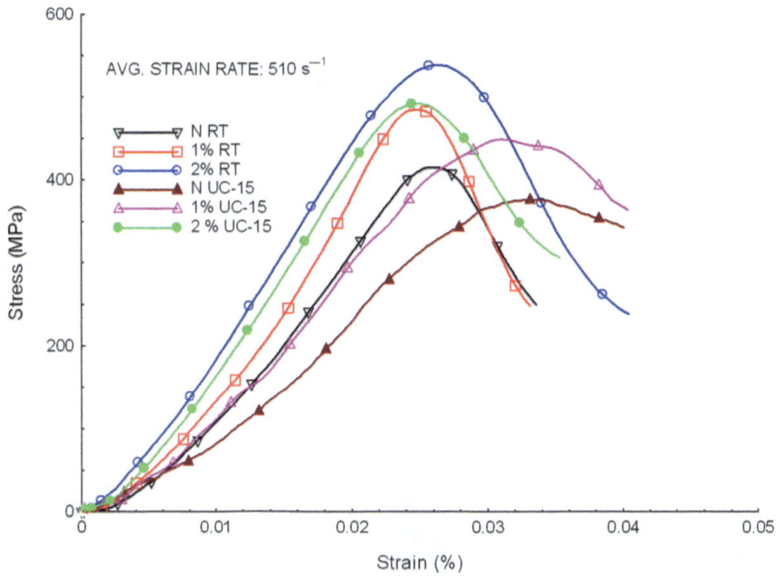

Fig. 5.12 Stress-strain curve of samples tested at baseline (room temperature) and after 15 days of UC conditioning at a strain rate of 510/s

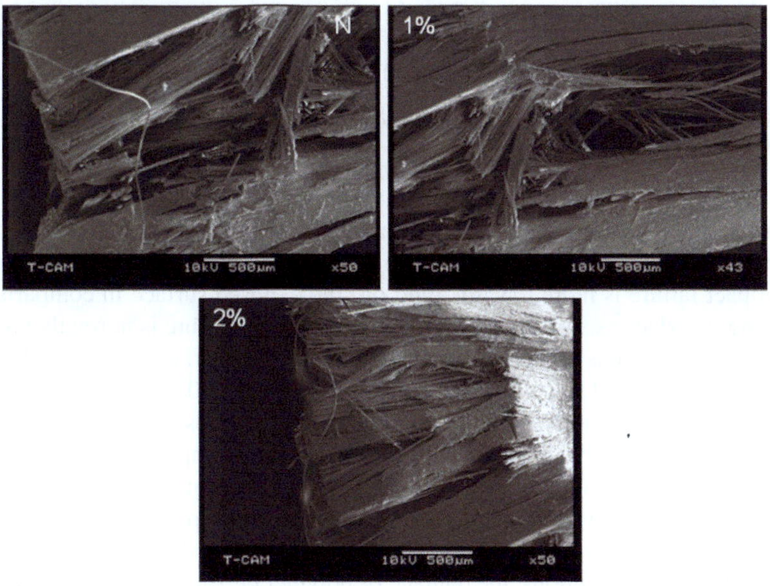

Fig. 5.13 Fracture behavior of neat, 1 wt% and 2 wt% samples subjected to UC conditioning for 15 days and tested at a strain rate of 382/s

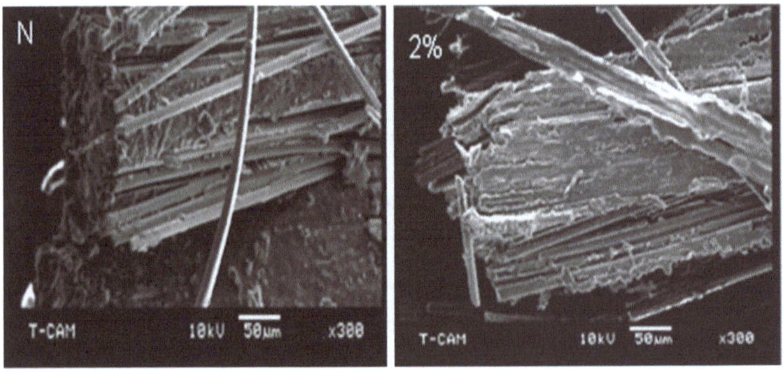

Fig. 5.14 Fracture behavior of neat and 2 wt% samples subjected to UC conditioning for 15 days and tested at a strain rate of 382/s

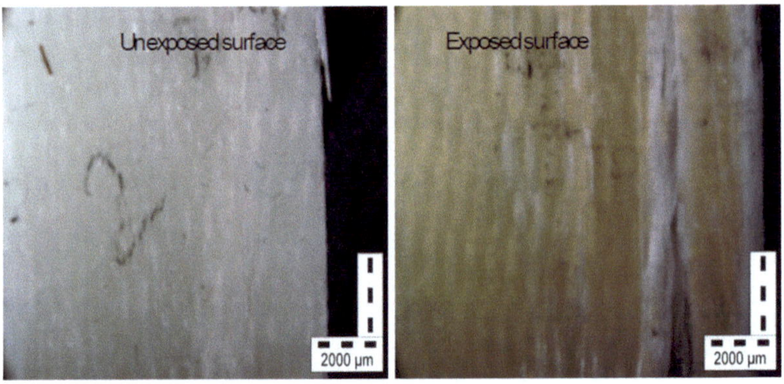

Fig. 5.15 Fracture surface of neat sample subjected to UV conditioning for 15 days and tested at a strain rate of 10^{-4}/s

Enhanced interfacial bonding between fiber and matrix was observed even after the failure of sample. In addition, it can also be observed from these micrographs that the impact failure is more towards (and on) the exposed surface in comparison to unexposed surface as shown in Fig. 5.15. Ultraviolet exposure is normally confined to the top few microns of the surface as reported by Karbhari et al. (2003). However, all the RT, UV, and UC conditioned samples tested at 440/s and 510/s strain rate failed due crushing and no fiber/matrix de-bonding was observed. No considerable differences were observed in neat and nanoclay infused samples.

However, it is worth noting that all the UV and UC samples conditioned up to 5 days showed insignificant decreases in compressive properties in comparison to room temperature samples at all strain rates. When subjected to ultraviolet radiation and condensation conditions, the damage of polymer depends on the exposure time. If the exposure time is short, the damages may be negligible or reversible, as reported by Regel et al. (1967). In contrast, Woo et al. (2008) reported a decrease in tensile strength after 250 h and an increase afterwards up to 1,000 h of UV exposure. Similar results are found in this study with a decrease in strength and modulus in both UV and UC conditioned for 15 days (360 h) samples in comparison to room temperature samples at all strain rates as shown in Tables 5.1 and 5.2.

The decreasing trend in the compressive properties upon UV and UC exposure may be attributed to the chain scission and the chain crosslinking phenomenon. Chain scission lowers the molecular weight and strength of polymers, while chain crosslinking embrittles polymers, resulting in microcracking but would also increase the strength as noted by Kumar et al. (2002) and Rabek (1995). In the presence of ultraviolet radiation, the chain scission dominates due to the creation of free radicals in the presence of UV photons and oxygen. However, the addition of nanoclay that accelerates the curing enhances the chain crosslinking and provides barrier to UV radiation, which minimizes the degradation and thus results in the improvement and lesser degradation in the properties of 2 wt% samples in comparison to neat samples. Decrease in compressive properties of UC conditioned samples was found to be higher in comparison to only UV conditioned samples. This can be attributed to the combined effect of ultraviolet radiation and condensation. The first form of damage in FRP composite is usually matrix microcracking. These microcracks cause degradation in properties of FRP composites and also act as precursor to other forms of damage leading to failure (Nairn 2000). Ultraviolet radiation and temperature produce surface erosion and microcracks on the surface that provide a pathway for rapid ingress of moisture into the polymer. The deteriotion of polymer matrix alone is sufficient to cause degradation in structural performance and reliability (Kumar et al. 2002). In fiber reinforced polymers (FRPs), water diffused into the composites ends up either in the matrix or at the interphase region. Thus, the amount of moisture absorbed by matrix is significantly different than that by the fiber. This results in a mismatch in moisture-induced volumetric expansion between the matrix and the fibers, which leads to the evolution of localized stress and strain fields in composites by Lee and Peppas (1993). In the matrix, water acts as plasticizer, increases free volume, lowers glass transition temperature, and relieves the internal stress that is built up during processing of the composites (Magid et al. 2005). Moisture wicking along the fiber-matrix interface can degrade the interfacial bond, resulting in the loss of micro-structural integrity.

5.6 Conclusion

Glass/epoxy composites are sensitive to environmental conditioning especially under UV and condensation conditions. Weight loss was observed in UV condition samples, whereas UV + condensation (UC) samples showed increase in weight. Discoloration on the surface was noticed in all the conditioned samples. Addition of 1–2 wt% nanoclay decreased the weight loss/gain and less color change was observed in comparison to neat samples, similarly conditioned. Compressive properties were found to degrade with increase in time. 2 wt% epoxy composites showed enhancement in properties in all the conditions over neat composites. Optical and scanning electron micrographs showed better interfacial bonding in 2 wt% samples and less delamination in comparison to neat and 1 wt% samples when tested at the same strain rates. Optical micrographs also revealed damage on the UV exposed surface in static test, whereas little to no damage was observed on the unexposed surface. The intensity/extent of damage was also found to be higher in neat samples in comparison to 2 wt% samples. SEM micrographs showed surface erosion and resin spallation from surface exposed on all the UV and UC conditioned samples and resin was spalled from the surface of samples. Additionally, small pores and whitening on the exposed surface were observed in UC condition samples. SEM micrographs also revealed better interfacial bonding in 2 wt% samples in comparison to neat and 1 wt% samples.

Acknowledgements Authors would like to thank the financial support provided by U. S. Army Engineer Research Development Center-Construction Engineering Research Laboratory (ERDC-CERL).

References

Alexandre M, Dubois P (2000) Polymer-layered silicate nanocomposites: preparation, properties and uses of a new class of materials. Mater Sci Eng 28:1–63

ASTM D 695–02 (2002) Standard test method for compressive properties of rigid plastics, ASTM International, 7pages

Bonniau P, Bunsell AR (1981) A comparative study of water absorption theories applied to glass epoxy composites. J Compos Mater 15:272–293

Carrasco F, Pages P (2008) Thermal degradation and stability of epoxy nanocomposites: influence of montmorillonite content and cure temperature. Polym Degrad Stab 98:1000–1007

Fukushima Y, Inagaki S (1987) Synthesis of an intercalated compound of montmorillonite and 6-polyamide. J Incl Phenom 5:473–482

Giori C, Yamauchi T (1984) Effects of ultraviolet and electron radiations on graphite- reinforced polysulfone and epoxy resins. J Appl Polym Sci 29:237

Hackman I, Hollaway L (2006) Epoxy-layered silicate nanocomposites in civil engineering. Compos A Appl Sci Manuf 37:1161–1170

Haliwelll SM (1992) Weathering of polymers. RAPRA review reports: ISSN 0889–3144, 53, 5(5)

Karbhari VM, Chin JW, Hunston D, Benmokrane B, Juska T, Morgan R, Lesko JJ, Sorathia U, Reynaud D (2003) Durability gap analysis for fiber-reinforced polymer composites in civil infrastructure. J Compos Constr 7(3):238–247

Kumar BG, Raman PS, Toshio N (2002) Degradation of carbon fiber- reinforced epoxy composites by ultraviolet radiation and condensation. J Compos Mater 36(24):2713–2733

LeBaron PC, Wang Z, Pinnavaia TJ (1999) Polymer-layered silicate nanocomposites: an overview. Appl Clay Sci 15:11–29

Lee MC, Peppas NA (1993) Models of moisture transport and moisture induced stress in epoxy-composites. J Compos Mater 27:1146

Liau WB, Tseng FP (1998) The effect of long-term ultraviolet light irradiation on polymer matrix composites. Polym Compos 19:440–445

Magid BA, Ziaee S, Gass K, Schneider M (2005) The combined effects of load, moisture and temperature on the properties of E-glass, epoxy composites. Compos Struct 71:320–326

Malvar LJ (1996) Literature review of durability of composites in reinforced concrete. Special publications SP2008SHR. Naval Facilities Engineering Service Center, Port Hueneme, p 26

Maul P (2005) Barrier enhancements using additives. In: Fillers, pigments and additives for plastics in packaging applications, Pira International Conference, Brussels

Meyers MA (1994) Dynamic behavior of materials. Wiley, New York, Chapter 12

Nairn A (1993) Fiberreinforcedplastic (FRP) reinforcement for concrete structures: properties and applications. Developments in Civil Engineering, Elsevier Science 42:450

Nairn JA (2000) Matrix microcracking in composites. Compr Compos Mater 2(12):403–432

Rabek JF (1995) Polymer photo-degradation: mechanism and experimental methods. Chapmen & Hall, London, pp 24–66, Chapter 2 & 10

Regel VR, Chernyi NN, Kryzhanovskii VG, Boboev TB (1967) Effect of ultraviolet radiation on the creep rate of polymers. Mekhanika Polimerov 3:615–618

Ricky SC, Woo C, Yanghai Z, Honggang L, Jing K, Jang-Kyo KY, Christopher Leung (2007) Environmental degradation of epoxy–organoclay nanocomposites due to UV exposure. Part I: photodegradation. J Compos Sci Technol 67:3448–3456

Tjandraatmadja GF, Burn LS, Jollands MC (2002) Evaluation of commercial polycarbonate optical properties after QUV-A radiation-the role of humidity in photodegradation. Polym Degrad Stab 78:435–448

Uhl FM, Davuluri SP, Wong SC, Webster DC (2004) Organically modified montmorillonites in UV curable urethane acrylate films. Polymers 45(18):6175–6187

Usuki Y, Kojima M, Kawasumi A, Okada Y, Fukushima TK (1993) Synthesis of nylon 6–clay hybrid. J Mater Res 8(5):1179–1184

Weitsman YJ, Guo YJ (2002) Correlation between fluidinduced damage and anomalous fluid sorption in polymeric composites. Compos Sci Technol 62:889–908

Woo RSC, Zhu HG, Leung CYK, Kim JK (2008) Environmental degradation of epoxy–organoclay nanocomposites due to UV exposure: part II residual mechanical properties. J Compos Sci Technol 68:2149–2155

Yuanxin Z, Farhana P, Mohammad AB, Vijaya KR, Jeelani S (2006) Fabrication and characterization of montmorillonite clay-filled SC-15 epoxy. Mater Lett 60:869–873

Zainuddin S, Hosur MV, Zhou Y, Ashok K, Jeelani S (2009) Durability studies of montmorillonite clay filled epoxy composites under different environmental conditions. Mater Sci Eng A 507:117–123

Chapter 6
FRP Reinforced Concrete Exposed to Elevated Temperatures: The Role of Viscosity on Bond Strength

Mohamed A. Faruqi, Hector Estrada, Claudia Gonzalez-Liendo, and Joseph O. Sai

Abstract There is overwhelming consensus among researchers that the bond strength between concrete and fiber reinforced polymer composite materials would be the first to yield when the composite is exposed to high temperatures. In this chapter, a semi-empirical model is developed that accounts for molecular bond breakup and viscosity conversion degree rate values to predict the normalized bond strength at elevated temperatures. The model is assessed with limited experimental data from available literature. Our results show that the proposed approach correlates well with the available experimental results. This formulation can be used for design of FRP reinforced concrete members at elevated temperatures, particularly in identifying critical bond strength and comparison of alternative proposed FRP reinforcing systems.

Keywords Bond strength • Elevated temperatures • Fiber reinforced polymers • Viscosity

M.A. Faruqi (✉) • J.O. Sai
Department of Civil and Architectural Engineering, Texas A & M University-Kingsville, Kingsville, TX 78363, USA
e-mail: m-faruqi@tamuk.edu; J-Sai@tamuk.edu

H. Estrada
Department of Civil Engineering, University of Pacific, 3601 Pacific Ave., Stockton, CA 95211, USA
e-mail: hestrada@pacific.edu

C. Gonzalez-Liendo
Texas Department of Transportation, 1817 Bob Bullock Loop, Laredo, TX 78040, USA
e-mail: CGONZALE@dot.state.tx.us

R. Jain and L. Lee (eds.), *Fiber Reinforced Polymer (FRP) Composites for Infrastructure Applications*, Strategies for Sustainability, DOI 10.1007/978-94-007-2357-3_6,

Nomenclature

a, d	Parameters governing the vertical location of the schematic model's curve along the y-axis
b, c	Parameters governing the horizontal location of the schematic model's curve along the x-axis, also *b* is further defined as a coefficient inversely proportional to a polymer's degree of cross-linking
C_r	Polymer degree of cross-linking
k_1	Coefficient accounting for the effect of bond loss at relatively low temperatures
k_2	Experimentally derived variable
T	Temperature at the surface of the rebar in degrees Celsius
T_g	Polymer glass transition temperature at the surface of the bar
T	Temperature at the surface of the rebar in degrees Celsius along x-axis
Y	Normalized bond strength in schematic model presentation
τ^*	Normalized bond strength
τ^*_r	Normalized residual bond strength
α_g	Conversion degree for glass transition
α_r	Conversion degree for leathery to rubbery transition

6.1 Introduction

Fiber Reinforced Polymer composites were developed to address the demand for materials that can offer multiple desirable properties not attainable by traditional homogeneous materials. FRP composites have found some acceptance in the infrastructure industry, both for repair/retrofit of structural members, as well as new construction. Though many issues have been resolved regarding the performance of composites in these types of applications, many more remain to be solved, particularly as these issues relate to durability. In a durability gap analysis of FRP composites, Karbhari et al. (2003) found that the main obstacle to widespread use of composite materials is the lack of data related to durability of FRP composites in civil infrastructure, including for extreme elevated temperature events, such as fires. Therefore, composites have not been fully utilized in many civil infrastructure applications because these materials have yet to prove their effectiveness at elevated temperatures, or because of the lack of data to support their effectiveness at elevated temperatures. As indicated in Table 6.1 by Karbhari et al. (2003), data for most structural components, particularly fire endurance of these components, is critical to have in order for composites to find widespread acceptance in the civil infrastructure design industry. The two main components that are of interest are the composite material and the composite/substrate interface (or bond); the latter being the issue investigated in this chapter.

Bond performance at elevated temperatures has only recently begun to be studied. Of particular interest is the material's structural response to fire exposure,

Table 6.1 Ranking of importance of data for fire effects (Karbhari et al. 2003)

Application area	Flame spread			Fire endurance			Smoke and toxicity			Heat release		
Level	A	B	C	A	B	C	A	B	C	A	B	C
Internal reinforcement												
Composite rebar[a]	–	–	–	3	3	–	–	–	–	–	–	–
External reinforcement												
Beams/slabs/columns[b]	3/5	3/3	–	5/5	5/5	1/1	1/5	1/3	–	3/5	3/3	–
Seismic retrofit												
Columns/piers/walls	3/5	3/3	–	5/5	3/3	1/1	1/5	1/3	–	3/5	3/3	–
Deck systems												
Conventional beams/girders	3/5	1/1	–	5/5	3/3	1/1	1/5	1/3	–	3/5	3/3	1/1
Integral/composite beams/girders	3/5	3/3	NS	5/5	5/5	NS	1/5	1/3	–	1/5	1/5	–
Structural elements												
Wall panels, profiles	3/5	3/3	NS	5/5	5/5	NS	1/5	1/3	–	1/5	1/5	–

Key: *A* composite level, *B* composite/substrate interface, *C* substrate
Ranking: *5* Critical, cannot go forward without it, *3* Important, *1* Good to have
Notes
[a]It is assumed that composite rebar will be embedded in a non-flammable concrete substrate. There are presently no adhesives connecting rebar to the concrete so flammability of the interface is not an issue. The important fire issue here is structural integrity during fire
[b]The effect of fire on external composite (combustible) wrap depends on whether the repair or upgrade is in a confined space (interior of a building, or parking garage, etc.). Thus, the first of two numbers in this category represent repair or upgrade situations on outside areas and second of the two numbers represent cases in confined spaces

which can be characterized by a constitutive relationship that regulates the material's mechanical behavior at elevated temperatures (Purkiss 2007). This relationship, however, has yet to be fully developed. Furthermore, there is much information available concerning FRP-reinforced concrete (internally and externally reinforced), but the area least studied is the performance of FRP-reinforced concrete materials at elevated temperatures, as indicated in Table 6.2 by Karbhari et al. (2003). As a result of this lack of data, leading sources on the subject are inclined to restrict the use of FRP systems for purposes of reinforcing and strengthening existing concrete infrastructure systems (ACI 2008). Also, because information on the performance of FRP composite materials at elevated temperatures is limited, current codes and design guidelines should note the potential issues with FRP materials at elevated temperatures. Therefore, before the civil infrastructure construction industry can fully take advantage of FRP-reinforced composite materials, these materials must be proven to work at elevated temperatures, and effective design guidelines must be developed.

It is difficult to generalize the mechanical and thermal behavior of FRP composite materials exposed to elevated temperatures for two reasons: first, the behavior of FRP materials depend on a large number of variables, such as fiber and matrix types,

Table 6.2 Ranking of availability of data for fire effects (Karbhari et al. 2003)

Application area	Flame spread			Fire endurance			Smoke and toxicity			Heat release		
Level	A	B	C	A	B	C	A	B	C	A	B	C
Internal reinforcement												
Composite rebar	5	5	5	5	5	5	5	5	5	5	5	5
External reinforcement												
Beams/slabs/columns	5	5	5	5	5	5	5	5	5	5	5	5
Seismic retrofit												
Columns/piers/walls	3/5	3/5	3/5	3/5	3/5	3/5	3/5	3/5	3/5	3/5	3/5	3/5
Deck systems												
Conventional beams/girders	3	3	3	3	3	3	3	3	3	3	3	3
Integral/composite beams/girders	5	5	–	5	5	–	5	5	–	5	5	–
Structural elements												
Wall panels, profiles	3	3	3	3	3	3	3	3	3	3	3	3

Key: *A* composite level, *B* composite/substrate interface, *C* substrate
Ranking: *1* Widely available and validated, *3* Sparse and/or questionable, *5* Not available

fiber volume ratio, modulus of elasticity of fibers and matrix materials; second, the wide variation in FRP formulations (Chowdhury et al. 2007). As the temperature of a composite material reaches the resin matrix's glass transition temperature, T_g (the temperature at which the shear modulus of the resin drops significantly, by an order of magnitude), the resin will begin to soften, relinquishing its ability to bond the fibers together and transfer stresses between fibers; consequently, it is not surprising that at elevated temperatures the bond between FRP materials and concrete would be the first to be impaired (Katz and Berman 2000) and (Katz et al. 1999).

Concrete members retrofitted with external FRP composite reinforcement must be capable of enduring fire safety requirements exclusive of any strength improvement contributed by the externally bonded system (Chowdhury et al. 2007). In internally reinforced concrete applications, the FRP material is protected against combustion because it is surrounded by concrete and, thus, combustion is not of concern. For internal reinforcement applications, bond strength between concrete and FRP is expected to yield first when the system is exposed to elevated temperatures, particularly in extreme events such as fire. This happens when elevated temperatures exceed the glass transition temperature of the polymer matrix material (ACI 2008). Consequently, there is a need to understand the bond strength of FRP embedded in concrete at elevated temperatures, the structural response of which can be characterized by a constitutive model of mechanical behavior at high temperatures. The model must account for molecular bond breakup and viscosity conversion degree rate values, which can be used to predict bond strength at elevated temperatures.

These models have not yet been fully developed for most FRP materials, but we need to understand the bond strength of FRP embedded in concrete exposed to elevated temperature if FRP reinforcing bars are to gain full acceptance in the

Picture 6.1 I-580/I-880 collapse caused by a tanker truck fire on April 29, 2007 (*The San Francisco Chronicle* website, accessed November 5, 2009)

civil infrastructure industry. FRP materials have already been used in new construction of bridges, and the possibility of bridges being exposed to fire is likely, as shown in Picture 6.1.

The main objective of the work presented in this chapter is to develop an improved semi-empirical model to predict the normalized bond strength at elevated temperatures accounting for molecular bond breakup and viscosity conversion degree rate values at elevated temperatures. The model is verified using experimental data obtained from available literature sources.

6.2 Proposed Approach

A limited amount of research has been dedicated to the study of FRP-reinforced concrete bond strength subjected to elevated temperatures. Katz and Berman (2000) conducted pullout tests to establish the bond strength between concrete and FRP bars when exposed to elevated temperatures. Bai et al. (2008) described the changes in mechanical properties a polymer experiences under elevated temperatures, specifically with respect to elastic modulus and the viscosity of the material. However, none of these approaches provide an analytical account for molecular bond breakup and viscosity conversion degree rate values to predict the normalized bond strength at elevated temperatures. In this study, we combined results from Katz and Berman (2000) and Bai et al. (2008) to account for molecular bond breakup and viscosity conversion degree rate values to predict the normalized bond strength at elevated temperatures.

6.2.1 Proposed Model

Presented below is a semi-empirical numerical model that evaluates the changes in bond strength as a specimen is exposed to increasing temperatures. The model is based on a hyperbolic tangent formulation as expressed by:

$$y = a\tanh\left[-b\left(x-c\right)\right]+d \tag{6.1}$$

The technical reason for choosing this formulation is the polymer behavior at elevated temperatures. At increasing temperatures, the polymer chains go through three main transitions that result in the breakup of molecular bond, namely: (1) glass, (2) leathery-to-rubbery, and (3) rubbery-to-decomposed. The model chosen captures this important physical phenomenon.

The equation is characterized along the y-axis by an upper and lower normalized bond strength (ɽ*); upper bound of $y = 1$ for a maximum bond strength representative at ambient temperatures and lower bound of $y = -1$ for a low bond strength representative of resin decomposition at elevated temperature. The *x*-axis represents the temperature at the surface of the rod, *T*, and the coefficients *a, b, c, d* account for bond strength at varying temperatures.

The viscosity of the polymer resin during the curing period has an influence on the cross-linking mechanism (ASM 1987). The degree of cross-linking plays a role in quantifying the bond strength between FRP bars and concrete when exposed to elevated temperatures (Katz and Berman 2000). For these reasons, we propose that the elements of molecular bond breakup and rate of conversion degrees be incorporated. The *b* value is indirectly a representation of rates ratio, over the molecular bond composition, which occurs as the polymer resin undergoes viscosity transitions. In other words, *b* incorporates the rate of molecular bond breakup and viscosity into the model; $b = k_2/C_r$. Where k_2 accounts for the conversion rate and is based on experimental data taken from reference Katz and Berman (2000), and C_r is the polymer cross linking related to viscosity. Also, k_1 accounts for bond loss. *b* is significant at glass transition temperatures lower than 80°C. The normalized bond strength, $ɽ^*_r$, is obtained by dividing residual bond strength at a particular temperature by the maximum bond strength at room temperature.

The numerical model that defines the normalized bond strength, ɽ*, is provided by:

$$ɽ^* = 0.5\left(1-ɽ^*_r\right)\tanh\left[-k_2/C_r\left(T-\left(T_g+k_1.C_r/k_2\right)\right)\right]+0.5\left(1+ɽ^*_r\right) \tag{6.2}$$

Where,

$$k_1 = 1, \quad for \quad T_g \le 80^\circ C,$$

$$1-0.025\left(T_g-80^\circ C\right) \quad for \quad 80^\circ C < T_g < 120^\circ C,$$

$$0, \quad for \quad T_g \le 120^\circ C$$

6.3 Calibration of Semi-empirical Model

Values of conversion degree rates for FRP concrete reinforcement used in the bond strength model were obtained from Bai et al. (2008). The evaluation was limited to CB (urethane modified vinyl ester resin core) and CPH (epoxy vinyl ester resin core) FRP reinforcing bars because the resin material had similar physical properties to the specimens tested in reference Katz and Berman (2000). Both types of bars are #4 average size, 12.7 mm nominal diameter. CB rebar had large deformations at the surface (a geometry similar to steel bars), with space between deformation of 7.5 mm; while CPH rebar had a narrow helical braid of glass fiber wound on them, with space between deformation of 27 mm (See Figs. 6.3 and 6.4). A thin sand coating was also applied to the CPH rebar to enhance bond strength. The T_g temperature for both CB and CPH bars were 124°C and 122°C, respectively. In assessing values for ratio k_2/C_r, best fit slope values were derived from Figs. 6.1 and 6.2, which were extracted from reference Bai et al. (2008). These slopes are related to the viscosity glass transition conversion degree (α_g) and the viscosity leathery-to-rubbery transition conversion degree (α_r) for CB and CPH bars. This is chosen due to rapid reduction of bond strength in the glass transition zone for CB bars and in leathery-to-rubbery zone for the CPH bars. The results of the analysis give the following values for the experimental coefficients accounting for the effect of bond loss, $k_2=0.009$ for CB bars and $k_2=0.0302$ for CPH bars. These results are used in the computation of the normalized bond strength.

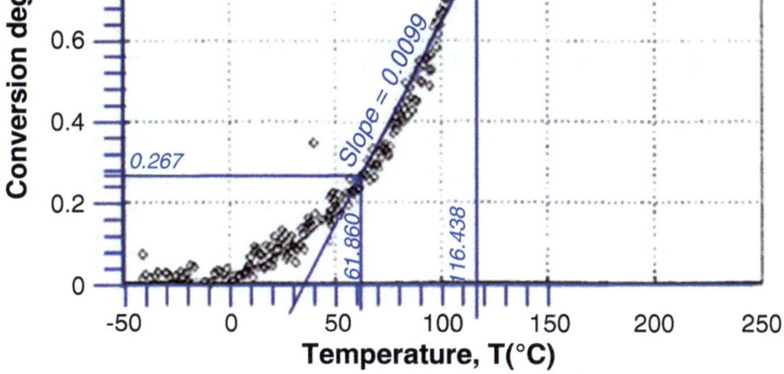

Fig. 6.1 Conversion degree of glass transition, α_g, for modeling viscosity (Bai et al. 2008)

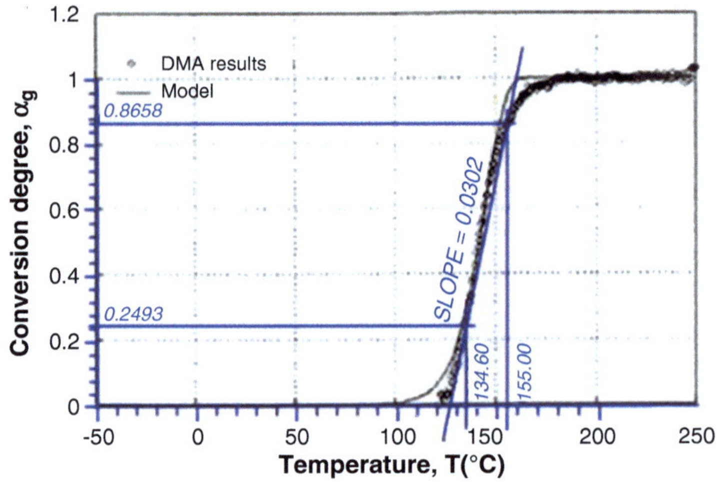

Fig. 6.2 Conversion degree of transition from leathery-to-rubbery state, α_r, modeling viscosity (Bai et al. 2008)

6.4 Calculations and Comparison of Normalized Bond Strength

The resulting normalized bond strength values from the proposed model are compared to experimental results. Figures 6.3 and 6.4 depict a comparison of normalized bond strength based on our semi-empirical model and the experimental data in reference Katz and Berman (2000). For the CB FRP reinforcing bars, temperatures ranging between 20°C and ~ T_g, the normalized bond strength values (glass transition, $k_2=0.009$) are compatible to the experimental results. Above T_g temperatures, the values are much closer to the experimentally measured bond strength. For the CPH FRP reinforcing bars, calculated normalized bond strength values (leathery-to-rubbery transition, $k_2=0.0302$) were closer to experimental bond strength measurements for all temperature ranges.

6.5 Concluding Remarks and Future Work

The work presented in this chapter is a semi-empirical investigation of bond strength of FRP composite reinforcing bars for concrete at elevated temperatures. The model presented accounts for molecular bond breakup and viscosity conversion degree rate. The model is assessed with limited experimental data from the available literature.

The following conclusions can be drawn from this work:

1. The model predicts the bond strength in CB composite reinforcement bars well for temperatures above T_g.
2. The model predicts the bond strength in CPH composite reinforcement well for all temperature ranges.

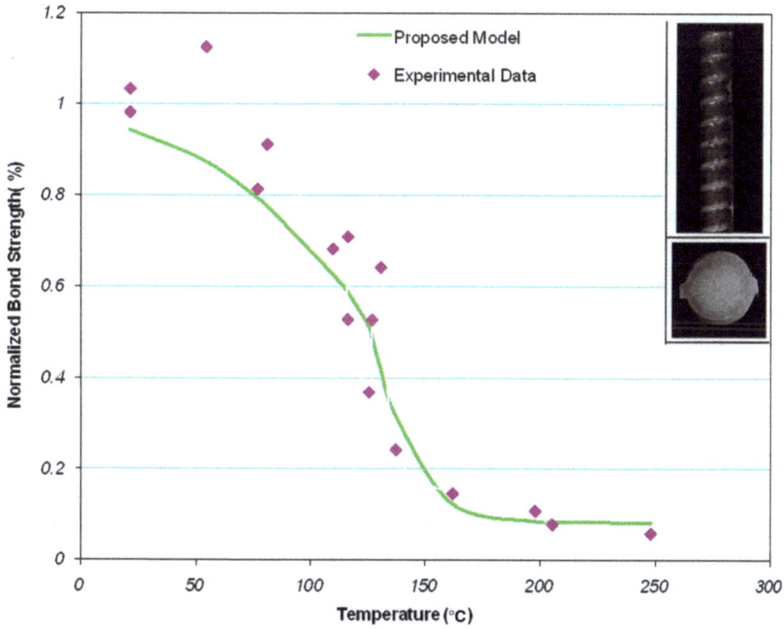

Fig. 6.3 CB rebar, comparison of normalized bond strength values (k_2=0.009); rebar: urethane modified vinyl ester as resin core

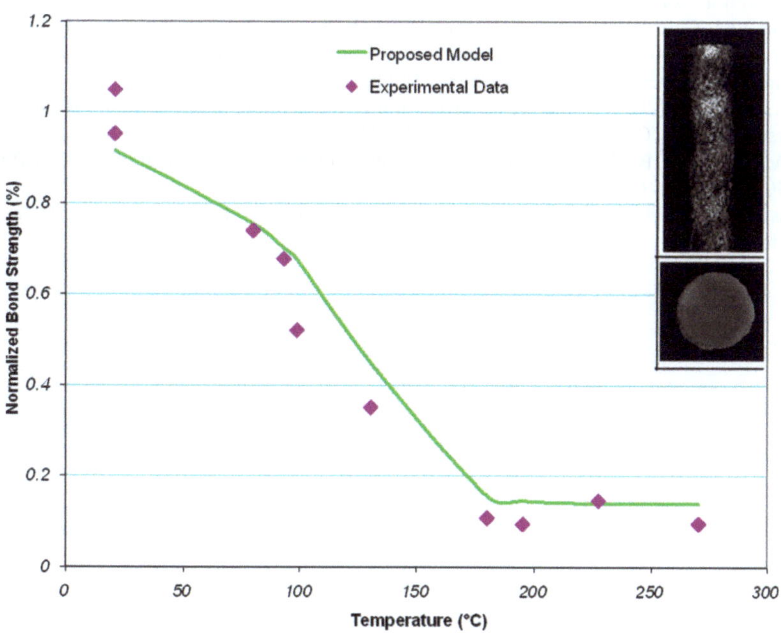

Fig. 6.4 CPH rebar, comparison of normalized bond strength values (k_2=0.0302); CPH rebar: epoxy vinyl ester as resin core

Further research is needed to verify the proposed semi-empirical model. Future research will include:

1. Other relevant properties of the rebar polymer surface incorporated in the model.
2. Better estimates of k_1 values based on glass transition temperature T_g.

Further research is needed to refine the semi-empirical bond strength model that will provide better accuracy in predicting FRP rebar behavior at elevated temperatures. Also needed is further collaboration to promote the acceptance of a practical design approach to temperature-dependence that is curtailing the use of FRP-reinforced concrete – a versatile, multi-property product.

References

American Concrete Institute (ACI) (2008) Report on fiber-reinforced polymer (FRP) reinforcement for concrete structures. ACI 440R-07, Farmington Hills

ASM International (1987) Engineered materials handbook composites, vol 1. ASM International, Metals Park

Bai Y, Keller T, Vallée T (2008) Modeling of stiffness of FRP composites under elevated and high temperatures. Compos Sci Technol 68:3099–3106

Chowdhury EU, Green MF, Bisby LA, Bénichou N, Kodur VKR (2007) Thermal and mechanical characterization of fibre reinforced polymers, concrete, steel, and insulation materials for use in numerical fire endurance modeling. National Research Council – Canada Institute for Research in Construction [NRC-CIRC] (NRCC-49684), Ottawa, Canada

Karbhari VM, Chin JW, Hunston D, Benmokrane B, Juska T, Morgan R, Lesko JJ, Sorathia U, Reynaud D (2003) Durability gap analysis for fiber-reinforced polymer composites in civil infrastructure. ASCE J Compos Constr 7(3):238–247

Katz A, Berman N (2000) Modeling the effect of high temperature on the bond of FRP reinforcing bars to concrete. Cem Concr Compos 22:433–443

Katz A, Berman N, Bank LC (1999) Effect of high temperature on bond strength of FRP rebars. ASCE J Compos Constr 3(2):73–81

Purkiss JA (2007) Fire safety engineering design of structures, 2nd edn. Butterworth-Heinemann, Oxford

Chapter 7
Impact Resistance of FRP Panels

Kiang Hwee Tan

Abstract Fiber reinforced polymer (FRP) panels may be used in deployable structures such as those required for the protection and mobility of armed forces. In this study, FRP panels were fabricated by hand in the laboratory and subjected to impact loading to investigate the resistance against impact actions. A striker with a hemispherical head having a diameter of 100 mm and weight of 43 kg, was allowed to drop freely along a frictionless guide from a height of 4 m onto the panels. Both carbon and glass FRP panels with up to ten plies of fiber sheets were investigated. The panels were subjected to a maximum of ten successive impacts. The impact characteristics, including the impact force and impulse, strains and displacements, and energy absorption capacity, are reported and compared. Test results showed that carbon FRP panels were likely to fail by punching shear with a small amount of fiber reinforcement. Otherwise, the energy absorption capacity of FRP panels increased with the amount of fiber reinforcement. Also, carbon FRP panels exhibited higher energy absorption capacity compared to glass FRP panels.

Keywords Energy absorption • Fiber reinforced polymer • Impact • Panels

7.1 Introduction

Fiber-reinforced polymer (FRP) composites have a low unit weight and relatively high strength, and are non-corrosive in nature, compared to traditional materials such as steel. In addition, they can be formed into various shapes even at room

K.H. Tan (✉)
Department of Civil & Environmental Engineering, National University of Singapore, Block E1A, #07-03, 1 Engineering Drive 2, Singapore 117576, Singapore
e-mail: cvetankh@nus.edu.sg

R. Jain and L. Lee (eds.), *Fiber Reinforced Polymer (FRP) Composites for Infrastructure Applications*, Strategies for Sustainability, DOI 10.1007/978-94-007-2357-3_7,

temperatures, and are, therefore, energy-saving and environmentally friendly. As such, FRP composites may be considered a sustainable material, with many potential infrastructure applications.

One application, based on the advantages of FRP composites, is their use in deployable structures for the armed services. As panels constitute one of the simplest forms of FRP structural components, its use in deployable structures is obvious. Deployable structures, such as small bridges and bunkers, however, may be subjected to impact loading.

Few studies have been carried out on the impact resistance of FRP composite panels with regard to their use in civil engineering. In a study on CFRP panels using a drop weight impact rig, Found and Howard (1995) concluded that failure of thin CFRP plates was dependent on the impact force rather than the energy. On the other hand, Choi et al. (1991) observed that impact damage occurred in graphite/epoxy laminated composites only when an impact energy threshold has been exceeded. Delamination with matrix cracking was observed during the initial impact damage. It was also found that the stacking sequence affected the impact resistance of the composites.

Belingardi and Vadori (2002) studied the impact characteristics of glass fiber composite plates and observed three broad categories of response: (1) rebound of impactor, (2) stoppage of impactor without penetration, and (3) perforation of specimens, regardless of the strain rate during impact. The force-displacement curves showed two threshold values corresponding to first damage and maximum force, and these were fairly constant when the impact energy was sufficiently large.

In another study, Zhou (1995) investigated the post-impact damage of thick glass fiber reinforced laminates using ultrasonic C-scanning and observed several predominant damage mechanisms, that is, delamination, fiber shear out, and fiber fracture. These mechanisms occurred sequentially with increasing incident kinetic energy or impact force and were also influenced by specimen geometry. The impact force levels required to initiate delamination increased with plate thickness, and load bearing capabilities were found to be higher for larger plates.

Aslan et al. (2003) also concluded from their study that specimen geometry plays a significant part on the contact force and duration. An analytical model developed using 3DIMPACT transient dynamic finite element analysis code and incorporating matrix cracking and delamination was found to predict the test results with fair reliability.

The present study was carried out to further investigate the impact resistance of carbon and glass FRP panels fabricated by the wet lay-up method. Laboratory impact tests were carried out using a drop-weight test rig to examine the impact characteristics of the panels. The number of plies of fiber sheets formed the main test parameter.

7.2 Test Program

The test program is summarized in Table 7.1. In all, six panels, comprising a group of three carbon FRP panels and another of three glass FRP panels, were fabricated. Each panel measured 1,000 mm by 1,000 mm on plan. For each group of FRP

Table 7.1 Test program

Specimen	Type of fiber reinforcement	# plies of fiber sheets	Average thickness (mm)	Fiber volume fraction (%)
PC4	Carbon fiber	4	3.45	19.1
PC6	Carbon fiber	6	5.75	17.2
PC10	Carbon fiber	10	10.1	16.3
PG4	Glass fiber	4	5.75	45.2
PG6	Glass fiber	6	6.95	56.1
PG10	Glass fiber	10	10.6	61.3

Table 7.2 Properties of carbon and glass fiber sheets

	Carbon fiber	Glass fiber
Fibre weight (g/m^2)	300	915
Fibre density (g/cm^3)	1.82	2.58
Thickness for design (mm)	0.165	0.353
Thickness after application of resin (mm)	0.45	0.65
Tensile strength (MPa)	3,500	1,700
Modulus of elasticity (GPa)	230	71
Maximum strain at rupture failure (%)	1.5	2.0

panels, the number of plies of fiber sheets was varied from four to six and ten. The sheets consisted of uni-directional fibers, with properties specified by the manufacturer, as in Table 7.2. The resin used to fabricate the FRP panels was supplied in two parts, that is, a main and a hardening agent, which were mixed in the proportion of 3:1 by weight, giving a blue paste.

7.2.1 Fabrication of Specimens

The steel template for making the panels was laid out first. Two plastic sheets, slightly larger than 1,000 mm by 1,000 mm, were cut. The face of each plastic sheet receiving the FRP panel was applied with wax to allow easy removal of the panel. One of the plastic sheets was placed onto the steel template (see Fig. 7.1a) with the waxed face up, and then two aluminum strips were adhered to the plastic sheet at 1,000 mm apart. A piece of Perspex was cut to form the third side, which also helped to align the fiber sheet.

The fiber sheets were cut to the required size of 1,000 mm by 1,000 mm. The first fiber sheet was placed onto the plastic sheet on the template and taped down on one edge. It was then flapped over the taped down edge. Resin was applied to the plastic sheet and the FRP sheet was flipped back onto the plastic sheet. Another coat of resin was applied as shown in Fig. 7.1b. The resin was brushed evenly to prevent air

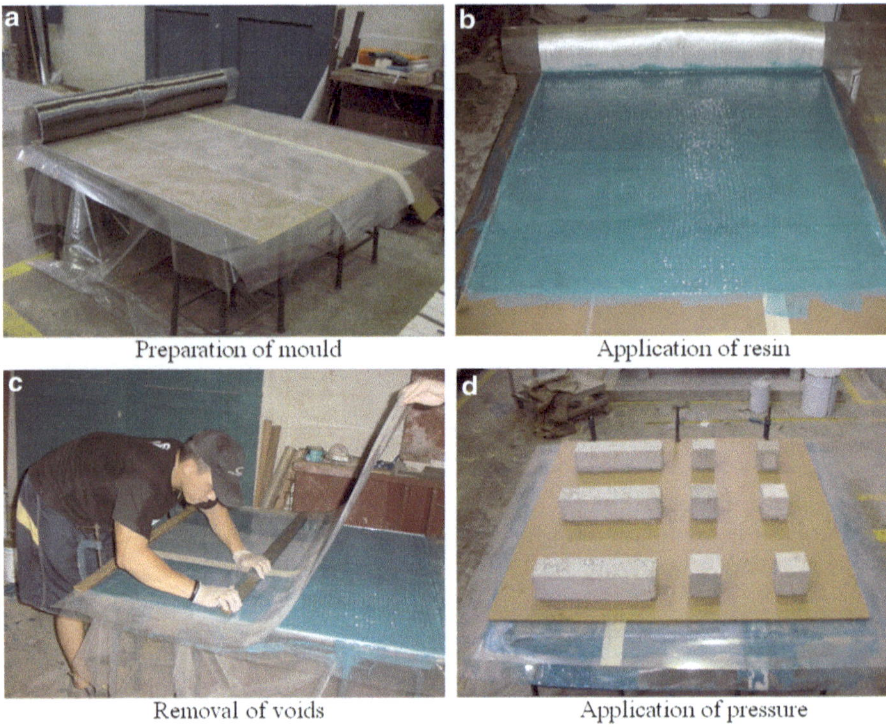

Fig. 7.1 Fabrication procedure

voids from forming, as shown in Fig. 7.1c. The second ply of fiber sheet was then laid in the orthogonal direction. The procedure was repeated until the required number of plies of fiber sheet was laid. The second plastic sheet was then flipped onto the composite panel.

After that, a large piece of Perspex board and some weights were placed on top of the panel to as shown in Fig. 7.1d. This step prevents formation of voids and ensures uniform thickness of the panel. The panel was left to dry and harden overnight and was removed from the template the following day. The fabricated FRP panels are shown in Fig. 7.2.

Mild steel plates measuring 900/1,000 mm in length, 50 mm in width, and 2 mm in thickness, were then bonded onto the perimeters of each panel on both faces using resin. This was done to facilitate the bolting and clamping of the panel onto the test rig.

7.2.2 Test Instrumentation and Set-Up

The strains at the bottom face of the FRP panels were measured by electrical strain gauges with a gauge length of 5 mm in two orthogonal directions, as shown in Fig. 7.3a. Three potentiometers were fixed to measure displacements, with two of

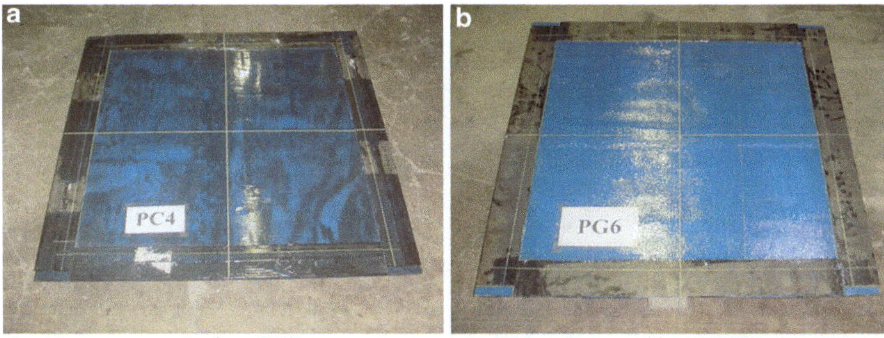

Carbon FRP Glass FRP

Fig. 7.2 Fabricated FRP panels

them placed at the center and the third at one third distance from the support line along the centerline of the panel.

An accelerometer was fixed at the tip of the hemispherical head of the striker to measure the deceleration and impact force. The diameter of the hemispherical head was 100 mm and the striker weighed 43 kg. Two laser diodes were also used to determine the impact velocity. The strain gauges, potentiometers, laser diodes, and accelerometer were connected to a 13-channel data acquisition system. The system was triggered by the striker passing the first laser diode. The recorded data has to be filtered in order to remove electrical noise. For this purpose, a cut-off frequency 3 kHz was selected by trial and error based on the Fast Fourier Transform (FFT) computation. The data was filtered digitally using the software Labview and based on second-order Butterworth filter.

Each panel was bolted and clamped onto the test rig on all four edges as shown in Fig. 7.3b. For each impact, the striker was raised to a height of 4 m and allowed to drop under its free weight onto the center of the panel. The line of fall was kept vertical by smooth rollers running along the aluminium guides. Grease was applied to the sides of the guide to reduce friction on the rollers. Each panel was subjected to a maximum of ten successive impacts.

7.3 Test Results and Discussion

Of the six FRP panels, Specimens PC4 and PC6 were both punched through at the first impact (see Fig. 7.4). The failure was catastrophic and the striker perforated the carbon fiber sheets. White patches were observed in the ruptured fiber sheets near the point of impact due to sudden increase in strains. Also, minor inter-ply delamination between the carbon fiber sheets was noted.

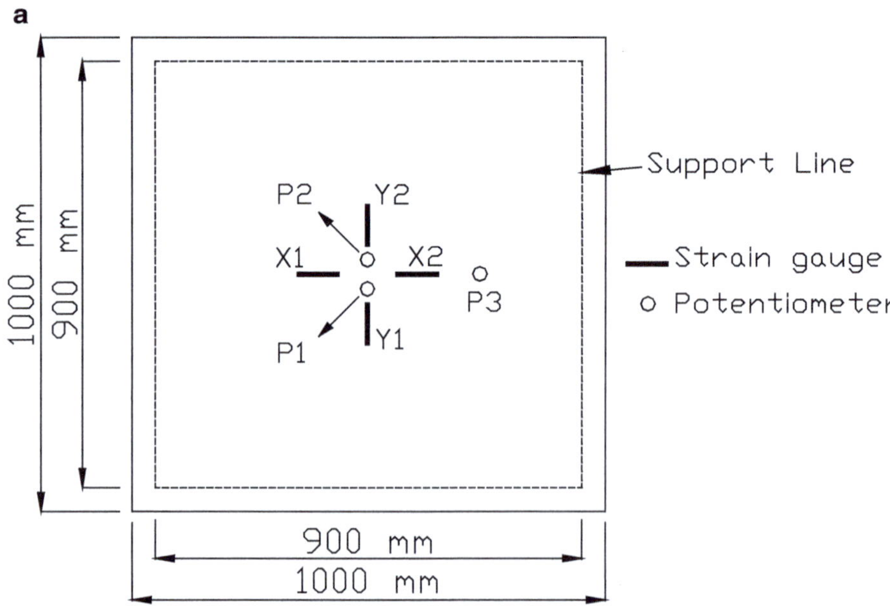

Locations of strain gauges and potentiometers

Test set-up

Fig. 7.3 Instrumentation and test set-up: (**a**) Location of strain gauges and potentiometers; (**b**) Test set-up

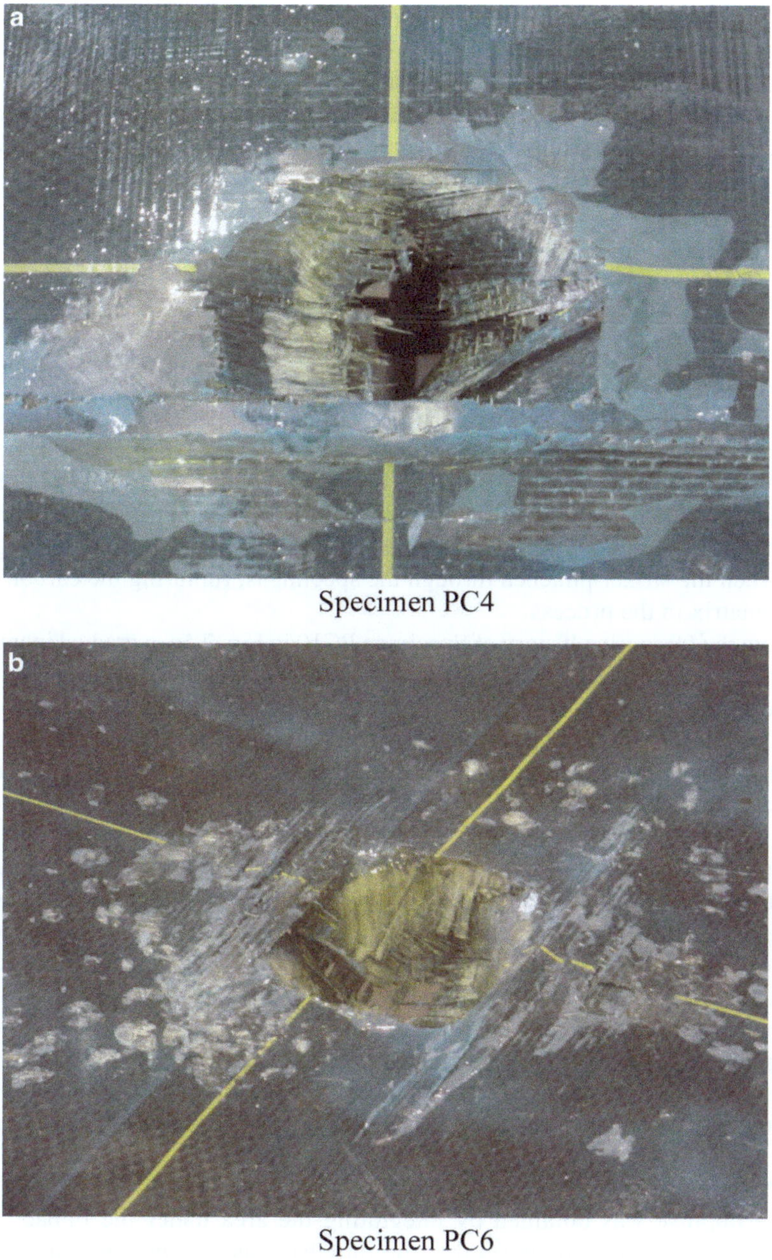

Fig. 7.4 Failure characteristics: (**a**) Specimen PC4; (**b**) Specimen PC6

Specimen PG4 was subjected to six impacts before the potentiometers were damaged and no further test data could be recorded. Slight damage was observed on the surface of the panel before the final impact. The remaining FRP panels, that is, Specimens PC10, PG6, and PG10, were each subjected to ten successive impacts. These four panels did not punch through and did not show any sign of damage on the surface.

7.3.1 Force-Time History and Impulse

The force-time history of the panels is shown in Fig. 7.5a–c for carbon FRP panels and Fig. 7.5d–f for glass FRP panels. For carbon FRP panels, at the first impact, the impact force was largest in Specimen PC6 at 110 kN, next in PC4 at about 80 kN, and smallest in PC10 at 55 kN. The high impact forces recorded in Specimens PC4 and PC6, however, cannot be totally relied upon as they were largely due to numerous secondary peaks that were the result of loud noise generated when the striker punched through the specimens, rupturing the carbon fibers in the matrix in the process.

From the force-time history of Specimen PC10 in Fig. 7.5c, a gradual increase in the impact force with the number of impacts was observed initially. This was probably due to the slack in the support restraint, which stabilized after two to three impacts. Thereafter, the impact force gradually decreased with successive impacts, probably due to initiation of damage. The impact force, however, became relatively constant after six successive impacts.

In the case of glass FRP panels (Fig. 7.5d–f), the impact force did not vary much with successive impacts. Some secondary peaks were observed. These could be attributed to the rebound of the striker and its re-contact with the panel. The impact force increased with the number of plies of glass fiber sheets at the first impact due to slack in support restraint. The duration of impact, however, seemed to decrease with increased plies of fiber sheets.

Comparing carbon FRP panels with glass FRP panels, we find that the carbon FRP panels are more susceptible to punching through failure due to the brittle properties of the fibers in the transverse direction. However, with sufficient number of plies of carbon fiber sheets, carbon FRP panels were able to resist the same impact force as glass FRP panels, as seen from the force-time histories of Specimens PC10 and PG10 in Fig. 7.5c–f respectively.

The impulse was obtained by integrating the area under the impact force-time history. Figure 7.6 shows that the impulse was between 500 and 600 Ns for most cases. The high impulse at the third impact for Specimen PC10 was probably an outlier.

a

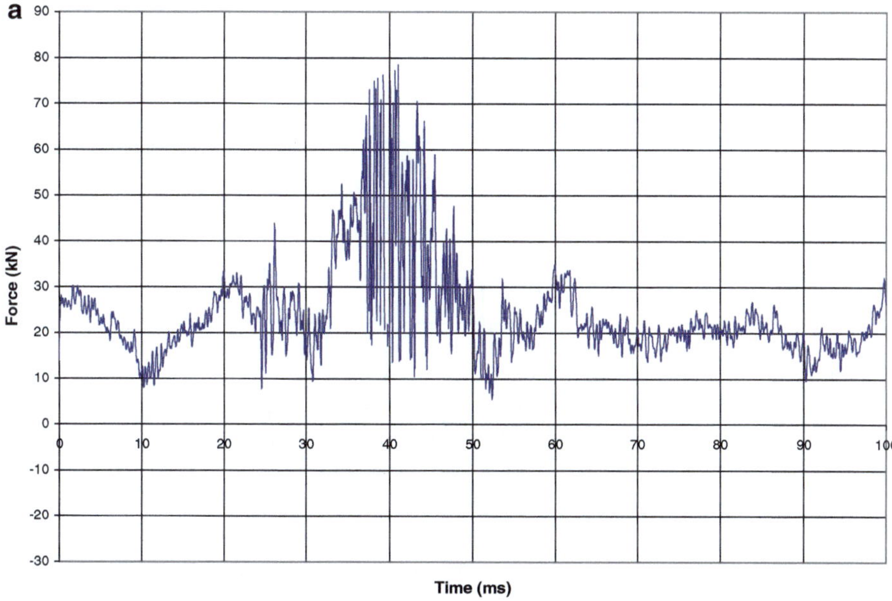

Specimen PC4

b

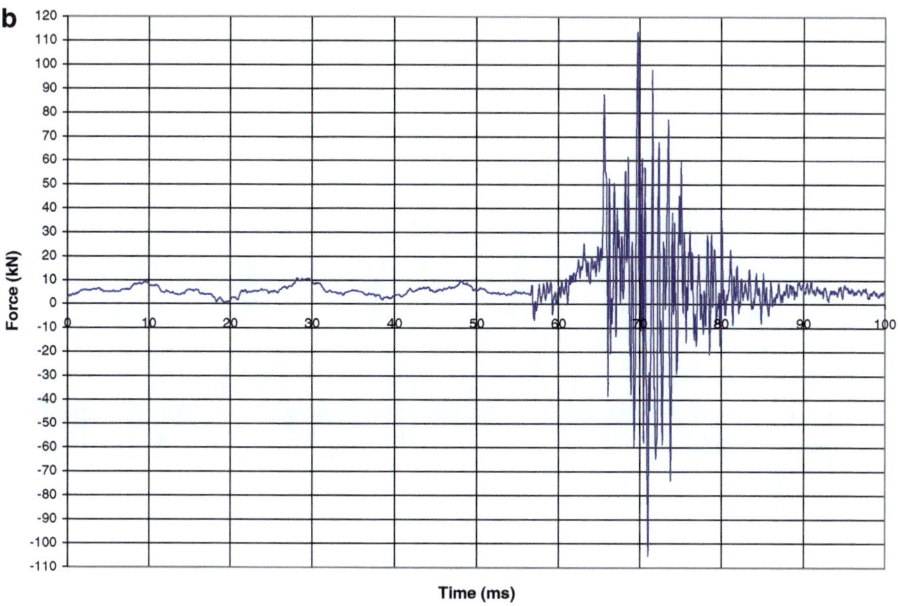

Specimen PC6

Fig. 7.5 Force-time histories: (**a**) Specimen PC4; (**b**) Specimen PGC6; (**c**) Specimen PC10; (**d**) Specimen PG4; (**e**) Specimen PG6; (**f**) Specimen PG10

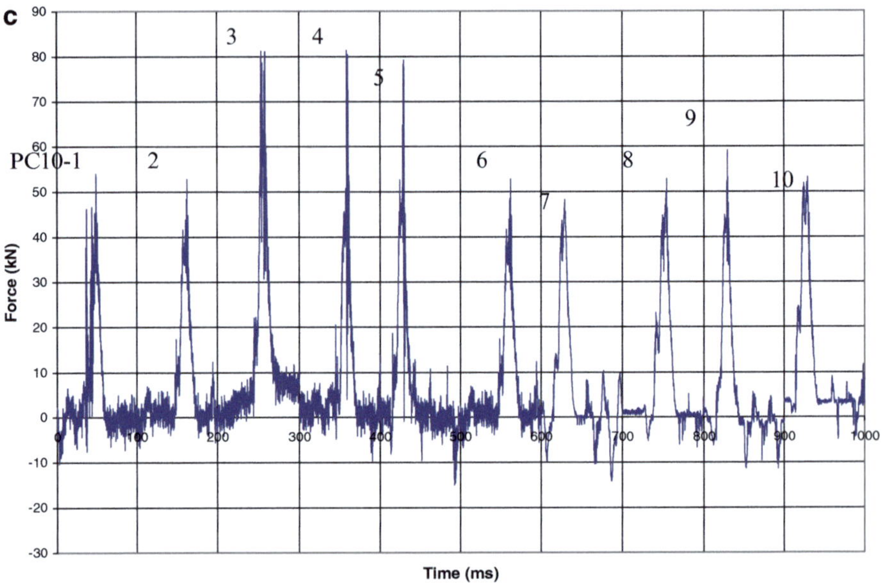

Specimen PC10

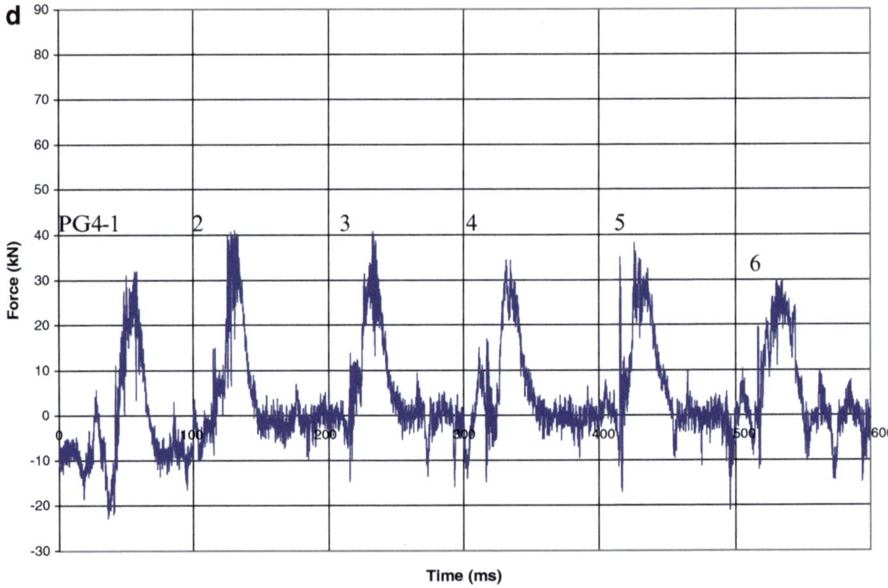

Specimen PG4

Fig. 7.5 (continued)

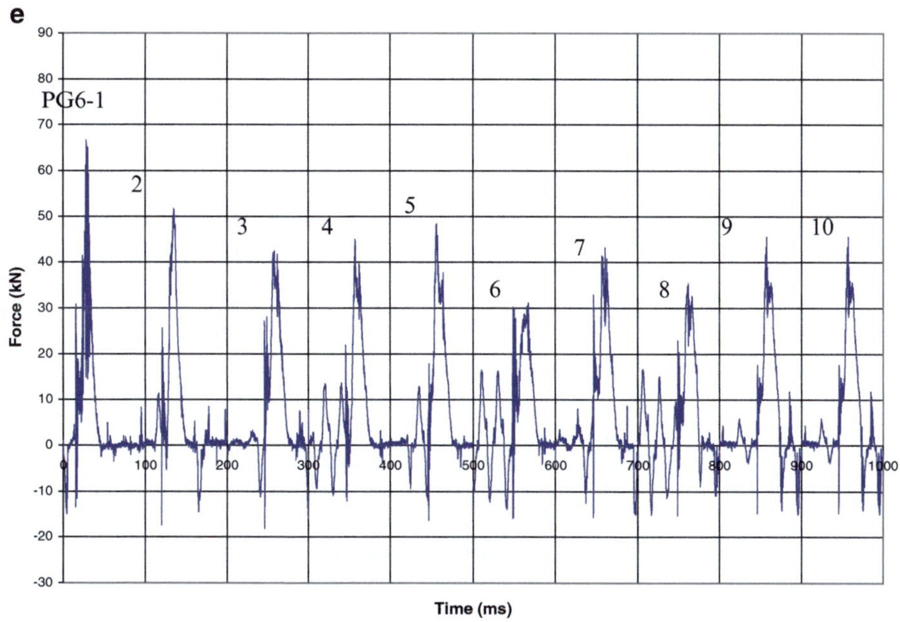

Specimen PG6

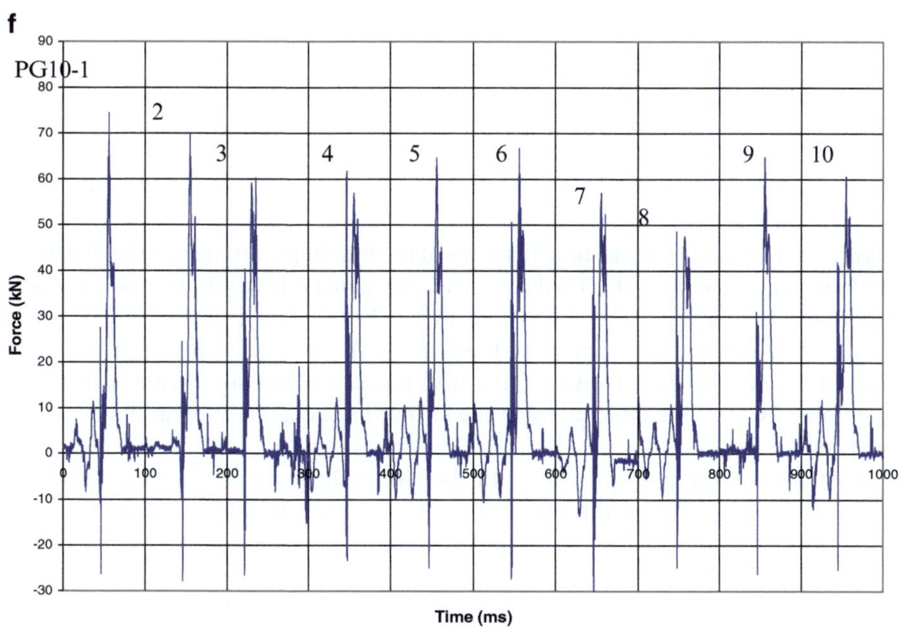

Specimen PG10

Fig. 7.5 (continued)

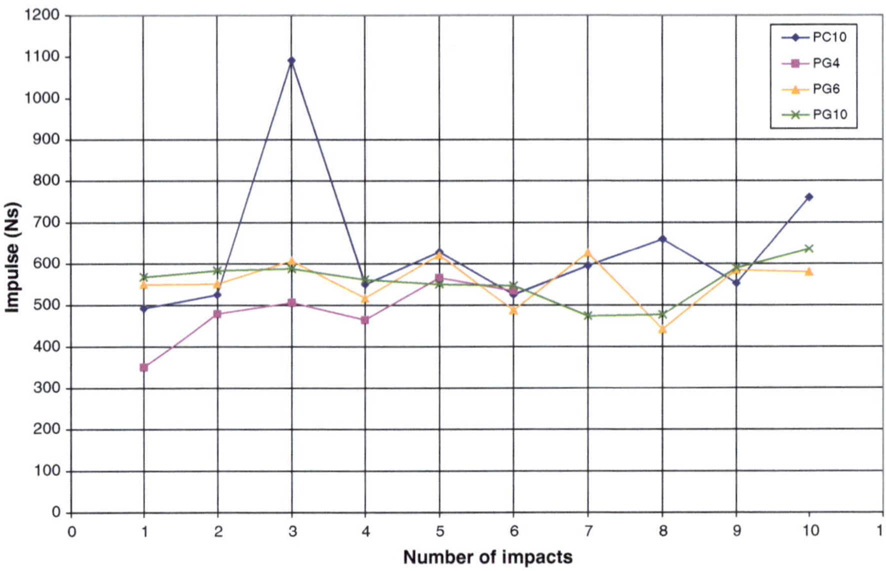

Fig. 7.6 Impulse

7.3.2 Strains and Displacements

Figure 7.7a shows the strain-time histories at the first impact for carbon FRP panels. A similar response was observed for glass FRP panels. It should be noted that the outermost bottom fiber sheet was laid in the Y-direction (see Fig. 7.3a) of the panel. Thus, the average strain readings in the Y-direction were slightly higher than in the X-direction. However, the difference was not significant, except for Specimen PC10. This indicated that the striker has accurately hit at the center of the panel.

Considering the average values in the two orthogonal directions, the maximum strain sustained by the panel at the first impact was found to be between 0.5% and 0.6%, regardless of the plies of fiber sheets. This was probably due to the predominance of membrane action at maximum deflection of the panel. However, Fig. 7.7b shows that the maximum strain increased gradually with successive impacts. This was due to slackening of the support clamps for the initial impact and to the laminates becoming compliant with increasing damage during subsequent impacts.

The displacement-time histories for Specimens CP10, PG4, PG6, and PG10 are plotted in Fig. 7.8, while the maximum displacement sustained by each panel is

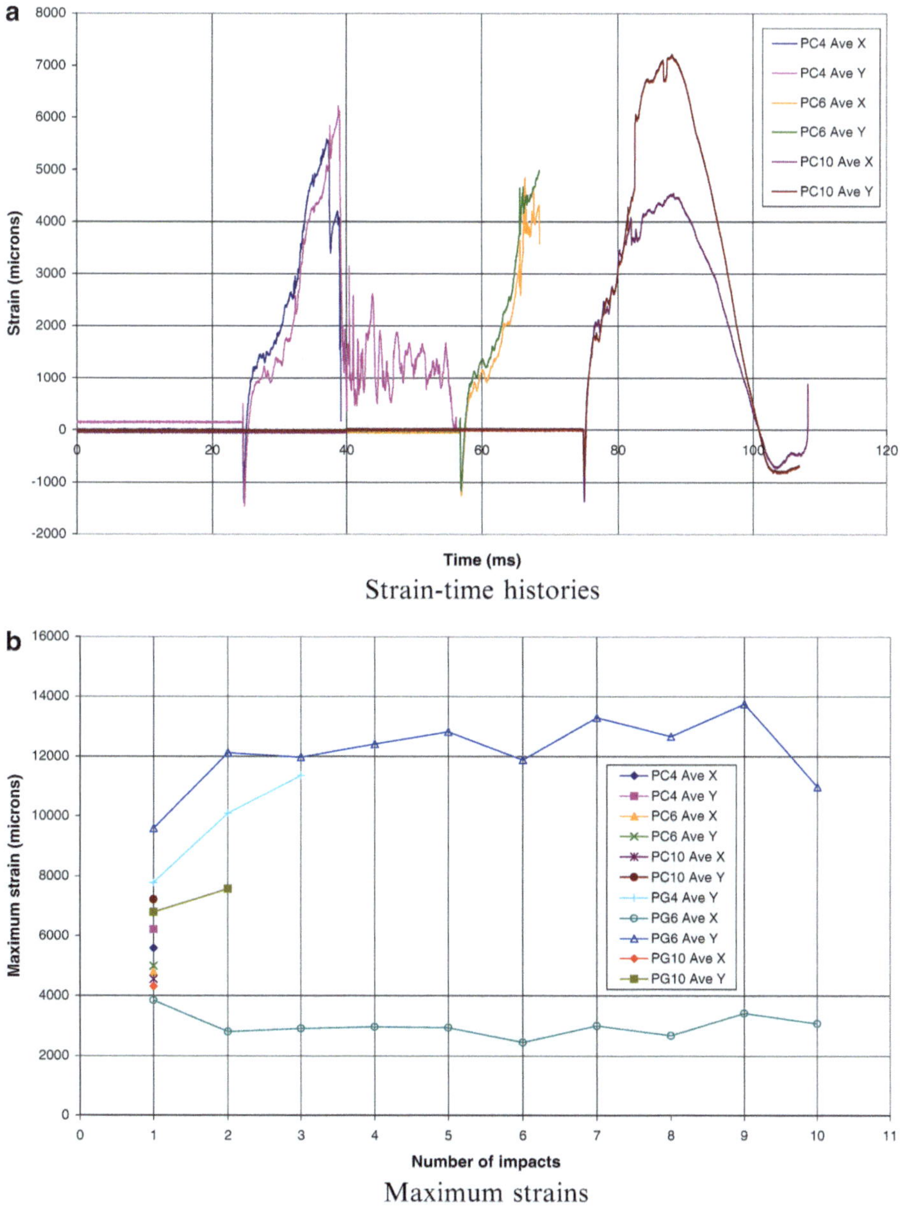

Fig. 7.7 Strains in FRP panels: (**a**) Strain-time histories; (**b**) Maximum strains

shown in Fig. 7.9. Readings for Specimens CP4 and CP6 were not available since they failed at the first impact. Overall, consistent with the strain readings, the maximum displacement did not vary significantly among the panels and increased only slightly if not remaining constant with successive impacts.

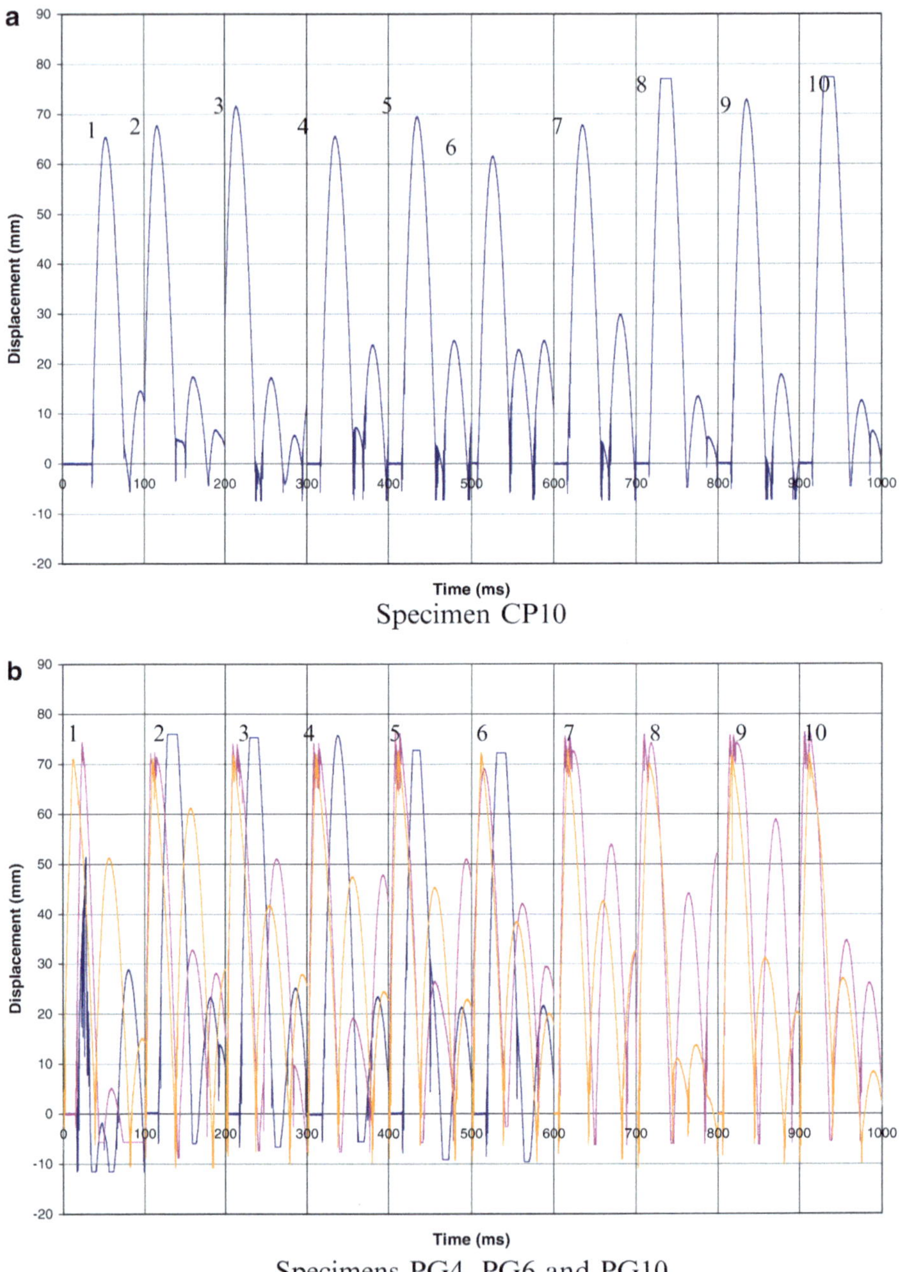

Fig. 7.8 Displacement-time histories: (**a**) Specimen PC10; (**b**) Glass FRP panels

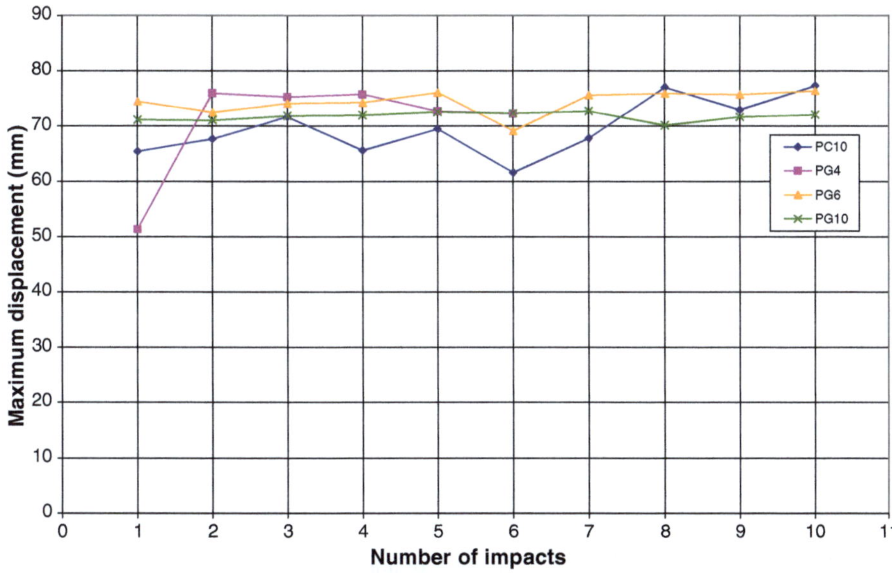

Fig. 7.9 Maximum displacements

7.3.3 Energy Absorption Capacity

The impact force-deflection relations were first obtained from the force-time and displacement-time histories. The energy absorbed by the panel during impact was then obtained as the area under the impact force-displacement curve. Figure 7.10a shows that the energy absorption by Specimen PG4 was not much different from PG6. However, the energy absorption capacity was greatly increased with ten plies of glass fiber sheets as in Specimen PG10. Compared to Specimen PG10, Specimen PC10 possessed higher energy absorption capacity. The energy absorption capacity, however, did not seem to decrease with successive impacts regardless of the type of FRP panels.

Figure 7.10b shows that the total (cumulative) energy absorbed by each panel increased almost linearly with successive impacts, indicating that the panels suffered little damage, if any. As it is with the energy absorbed at each impact, the cumulative energy absorption capacity of glass FRP panels increased only when the fiber sheets was increased beyond six plies. Also, the carbon FRP panels exhibited higher energy absorption capacity compared to glass FRP panels if they do not punch through under the impact load.

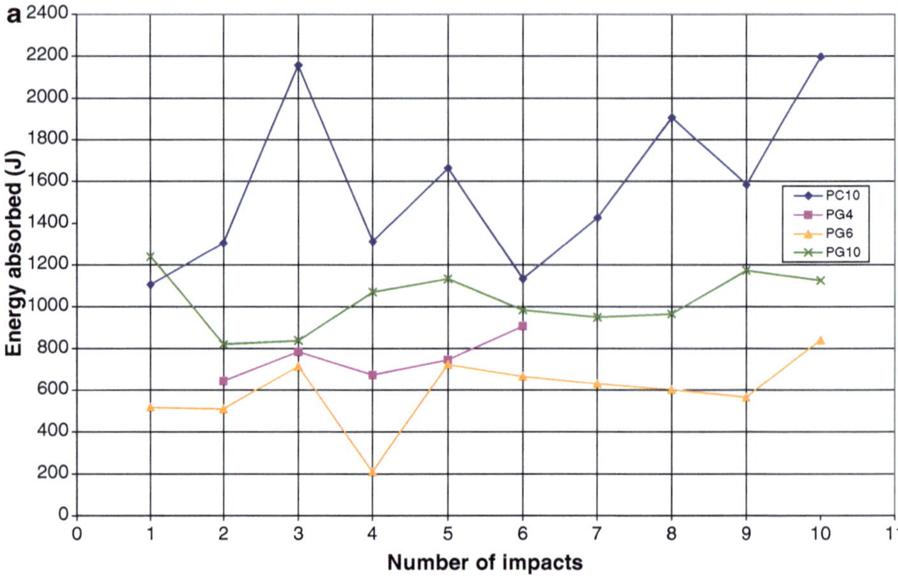

At each impact

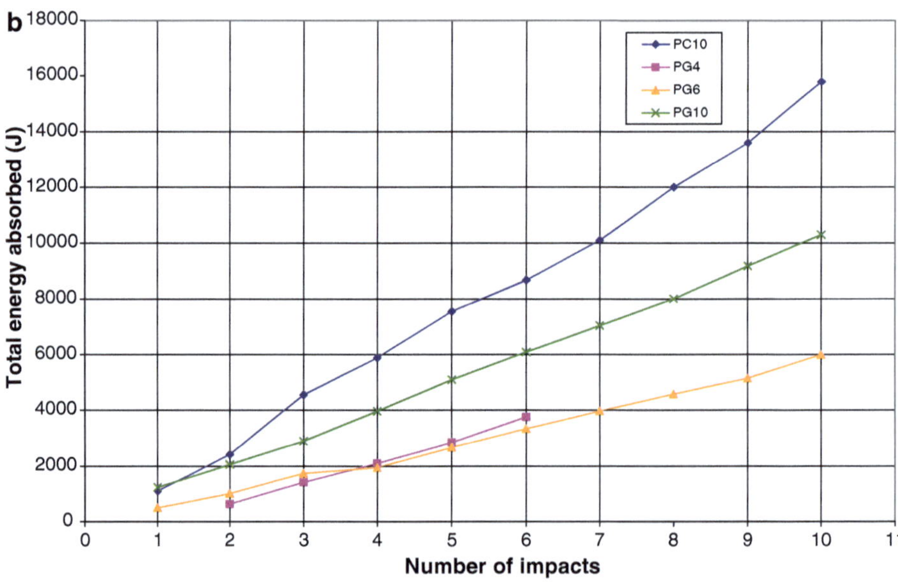

Cumulative

Fig. 7.10 Energy absorption capacity

7.4 Conclusions

From the study on carbon and glass FRP composite panels fabricated using four, six, and ten plies of fiber sheets and subjected to impact loading, the following conclusions may be made:

1. Carbon FRP panels with a small amount of fiber sheets were susceptible to punching shear failure due to weak transverse properties of the fiber sheets.
2. The energy absorption of FRP panels increased with the amount of the fiber reinforcement if punching shear failure is prevented.
3. Carbon FRP panels exhibited higher energy absorption capacity than glass FRP panels with the same number of fiber sheets, provided that they do not fail by punching.

References

Aslan Z, Karakuzu R, Okutan B (2003) The response of laminated composite plates under low-velocity impact loading. Compos Struct 59:119–127

Belingardi G, Vadori R (2002) Low velocity impact tests of laminate glass-fiber-epoxy matrix composite material plates. Int J Impact Eng 27:213–229

Choi HY, Downs RJ, Chang FK (1991) A new approach towards understanding damage mechanisms and mechanics of laminated composites due to low-velocity impact: part I – experiments. J Compos Mater 25:992–1011

Found MS, Howard IC (1995) Single and multiple impact behavior of a CFRP laminate. Compos Struct 32:159–163

Zhou G (1995) Damage mechanisms in composite laminates impacted by a flat-ended impactor. Compos Sci Technol 54:267–273

Part II
Applications

Chapter 8
Steel-Free Bridge Decks Reinforced with FRP Composites

Lijuan Cheng

Abstract Successful utilization of advanced polymer composite materials in civil infrastructures have long been recognized in bridge deck applications. The concept of 'steel-free' in concrete bridge decks can be achieved by replacing the conventional steel reinforcement with fiber reinforced polymer (FRP) based reinforcing materials. Reviewed in this chapter are steel-free bridge decks reinforced with these composite reinforcement in forms of round or square rods, 2-D or 3-D grids or gratings, flat or curved FRP plate, sandwich panel with foam/balsa/steel inserts, stiffened or corrugated stay-in-place formwork, and polymer decks.

Keywords Steel-free • Fiber reinforced polymer (FRP) • Stay-in-place (SIP) • Bridge decks

8.1 Introduction

Modern bridge construction is quickly evolving to the next generation where the structure itself is more corrosion-resistant and long-lasting, and the construction speed is greatly accelerated. The concept of material "steel-free" has already been introduced to critical components in concrete bridges such as decks in both design and construction stages. Advanced high-performance materials such as Fiber Reinforced Polymer (FRP) composites are no longer the secondary materials used for strengthening, rehabilitation and retrofitting of existing deficient bridges.

L. Cheng (✉)
Department of Civil and Environmental Engineering, University of California Davis,
One Shields Ave., Davis, CA 95616, USA
e-mail: dawcheng@ucdavis.edu

R. Jain and L. Lee (eds.), *Fiber Reinforced Polymer (FRP) Composites for Infrastructure Applications*, Strategies for Sustainability, DOI 10.1007/978-94-007-2357-3_8,

They are being used as viable primary structural materials for new construction of vehicle and pedestrian bridges in the civil engineering field.

As the advancement of manufacturing techniques, FRP composites (such as carbon FRP, glass FRP, aramid FRP) for new bridge deck construction exist in a wide range of commercial off the shelf and customized reinforcement products. They are currently available in the following formats for bridge deck construction:

- Round or square rod (nonprestressed and prestressed)
- Continuous flat or curved FRP plate
- Grid and grating (two-dimensional and three-dimensional)
- Sandwich panel with foam, balsa or steel inserts
- Stiffened or corrugated stay-in-place formwork
- Polymeric decking system (e.g., pultruded FRP module)

This chapter reviews the use of fiber reinforced polymer composites in bridge deck systems as primarily construction material either in pure FRP form or in combination of conventional concrete resulting in steel-free bridge deck systems. The design concept, structural performance in contrast to the conventional steel-reinforced concrete system, and current research and application status of each of these systems are discussed in this chapter.

8.2 FRP Rod Reinforced Concrete Deck

Composite materials made of fibers and polymers have high tensile strength and noncorrosive properties, making them the viable substitute for steel as structural reinforcement to achieve steel-free or corrosion-resistant concrete systems. They have been introduced to concrete structures for more than two decades as non-prestressed (e.g., Nanni 1993; Hassan et al. 2000; ACI 2006) and prestressed reinforcement (e.g., Grace 1999; ACI 2004). FRP reinforcing rods or bars are commercially produced through a manufacturing process known as pultrusion. They can be in conventional solid round, rectangular or hollow shape with a non-smooth surface deformation such as ribbed, sand-coated or wrapped and sand-coated in order to better bond to the concrete substrate (ACI 2006). Note that the following discussions are devoted to the concrete slab reinforced with non-prestressed FRP rods since they are currently more commonly used in the field.

8.2.1 Structural Performance

FRP rods are made of anisotropic material (essentially transversely isotropic) and therefore, unlike steel, they provide different physical and mechanical properties in the longitudinal fiber direction and the transverse direction. In addition, most FRP rods do not exhibit the elastic-plastic behavior that steel bars have (yielding before rupture).

Because of these inherent material characteristics, concrete slabs reinforced with FRP rods behave differently from the conventional steel-reinforced concrete slabs. In particular, (a) the ductile tension failure traditionally occurred in steel-reinforced concrete slabs may not be observed in the FRP rod-reinforced concrete slabs since most FRP rods rupture in a very brittle and sudden manner (Vijay and GangaRao 2001); (b) the stress redistribution is difficult to achieve in FRP rod-reinforced concrete slabs due to the linear-elastic behavior of the FRP material. As a result, the final failure of the structure can be more catastrophic; (c) the FRP rods are typically not used in compression zone of the concrete section. If they do exist, their contribution to the compressive capacity of the slab is not recommended to rely on (ACI 2006); (d) the deflection of glass FRP (GFRP) rod-reinforced concrete slab tends to be larger than the steel-reinforced slab with the equivalent reinforcement ratio. This is mainly due to the relatively low modulus of GFRP rods in comparison with steel bars; and (e) the FRP rod-reinforced concrete slab shows similar failure mode but better fatigue performance with longer fatigue life as compared to the steel-reinforced ones under fatigue loading condition (El-Ragaby et al. 2007; Carvelli et al. 2010).

8.2.2 Design Concept

The basis for the design of FRP rod-reinforced concrete deck is similar to the design philosophy for conventional steel-reinforced concrete slabs where the design is governed by limit states incorporating load and resistance factors. The American Concrete Institute publication ACI 440.1R-06 (ACI 2006), *Guide to the Design and Construction of Structural Concrete Reinforced with FRP Bars*, outlines these basic design guidelines which are essentially compatible with the ACI 318–05 (ACI 2005), *Building Code Requirements for Structural Concrete and Commentary*, and the AASHTO specifications for bridge slabs (AASHTO 2008), *Standard Specification for Highway Bridges*. This design methodology can also been found in other nation's standards such as the ones developed by the Canadian and Japanese associations (CSA 2002; JSCE 1997) for use of FRPs in civil infrastructures. The major differences in comparison with the typical design of conventional steel-reinforced concrete slabs are outlined in this section, ranging from the aspects of design assumptions, procedures for shear and serviceability, resistance factor, detailing, and environmental consideration.

Most of the *design assumptions* typically adopted for the design of steel-reinforced concrete slab are still valid in the cases of FRP rod-reinforced concrete slabs, such as plane section remains plane and no relative slippage between the reinforcement and concrete (MacGregor 1997). However, most FRP rods (e.g., GFRP and CFRP) are purely linear elastic materials which do not go through any yielding phase before the final failure. Therefore, the constitutive relationship between the stress and strain for the reinforcement has to be revised to reflect this linear elastic response up to failure, instead of the multi-stage ductile response in steel bars. The yield strength of the steel rebar also needs to be replaced by the ultimate strength of the FRP rods in the analysis of the deck section. In addition, the contribution to section's compressive strength from the FRP rods is very small compared with steel. Thus, if they actually exist in the compression zone of the concrete slab, their contribution shall be assumed to be negligible (ACI 2006).

The *design procedure* of a conventional steel rebar-reinforced concrete slab assumes that the concrete and steel rebars are sufficient to provide the adequate shear strength of the slab which typically does not have any shear reinforcement. Therefore, no procedure on checking the shear is required in the design (AASHTO 2008). This is, however, no longer true when designing the FRP rod-reinforced concrete slabs (Michaluk et al. 1998; El-Sayed et al. 2005) where the transverse shear and punching shear strengths must be checked after the flexural strength is designed for the slab. In other words, the amount of tensile reinforcement in the FRP rod-reinforced concrete slab can depend on the shear capacity of the slab. Moreover, check on serviceability conditions such as deflection and crack width is not required for steel-reinforced concrete slabs either. As a result, an extra design procedure must be added for FRP rod-reinforced concrete slabs for serviceability conditions which also include the creep rupture stress (ACI 2006).

Similar to steel-reinforced concrete design, *resistance factor* is also distinguished among the different types of failure modes in FRP rod-reinforced concrete slabs. The existing ACI guide provides the resistance factors for both flexural and shear designs (ACI 2006). The factors for calculating the ultimate flexural capacity are generally lower than those suggested for the steel-reinforced concrete design (ACI 2005) to account for the more brittleness of some of the failure modes.

Special attention shall also be given when *detailing* the designed cross-section and FRP reinforcement. Due to the anisotropic nature of the FRP reinforcing materials, large mismatch of coefficient of thermal expansion exists in the transverse direction in the FRP rods and the concrete. This can cause longitudinal splitting cracks in the concrete when subjected to temperature variations (Aiello et al. 1999). Research shows that this longitudinal cracking can also seriously degrade the bond capacity of the FRP rods (Sen et al. 1998; ACI 2006). Therefore, the concrete cover of the slab section needs to be carefully selected to account for this thermal effect and avoid such splitting cracking. Similarly to steel-reinforced concrete slabs, the FRP reinforcing rods have to be detailed in terms of their position, development length, splices, etc, following the guidelines (ACI 2006). However, two unique facts should be reminded during the reinforcement detailing of FRP rod-reinforced concrete slab: (a) FRP rods with bends can only be achieved during their manufacturing stage and cannot be bent after the reinforcing products are shipped to the construction site; and (b) the entire design process can be iterative when the calculated amount of FRP rods from the earlier flexural design cannot provide sufficient strength to satisfy the detailing requirement on bond strength (Bank 2006).

The existing research data on durability of FRP rods show that although FRP rods are less susceptible to corrosion than steel bars, their mechanical properties can deteriorate under harsh environmental attacks (Karbhari 2007). *Environmental reduction factors* are thus specified in the ACI 440.1R-06 for various fiber types and exposure conditions in the design (ACI 2006).

8.2.3 Application Status

Extensive amount of experimental and theoretical research has been performed in exploiting the use of FRP rods as internal reinforcement for concrete slabs in replace of steel (e.g., Bank et al. 1998; Hassan et al. 2000; Benmokrane et al. 2004). These promising research progresses have led to the successful development of design guidelines (e.g., ACI 2006). In the recent years, FRP rods have been widely used in bridge field projects in North America where conventional steel bars are completely eliminated from the concrete bridge deck. Representative examples include the Pierce Street Bridge in Lima, Ohio, where FRP bars were effectively used to prevent salt-induced corrosion in the bridge deck (Balendran et al. 2002); the Walters Street Bridge in central Missouri, the concrete bridge in St James, Missouri, the Headingley Bridge in Manitoba, Canada (Rizkalla 1997), all of which uses both CFRP and GFRP reinforcement; and the bridge at 53rd Ave. in Bettendorf, Iowa (Nanni 2001).

8.3 FRP Grid or Grating Reinforced Concrete Deck

The conventional on-site construction of concrete deck requires labor intensive assembly and tying of individual reinforcing bars. FRP grids or gratings made of two-dimensional or three-dimensional prefabricated lattice structure in single or double layer provide a practical solution to facilitate and accelerate the construction process. Concrete bridge decks reinforced by these systems are typically referred as FRP grid or grating reinforced concrete deck, as illustrated in Fig. 8.1. The FRP grid typically consists of orthogonally intersecting longitudinal and transverse members that are composed of continuous reinforcing fibers, such as glass, carbon, aramid and hybrid carbon/glass, impregnated with appropriate resin systems. The commercially available thin flat or curved or three-dimensional NEFMAC grid (New Fiber Composite Material for Reinforcing Concrete) gives a representative example of such grid systems (Fujisaki et al. 1993; Autocon Composites 2010). A pin-winding process similar to filament winding is typically used to fabricate the grid with intersecting layers.

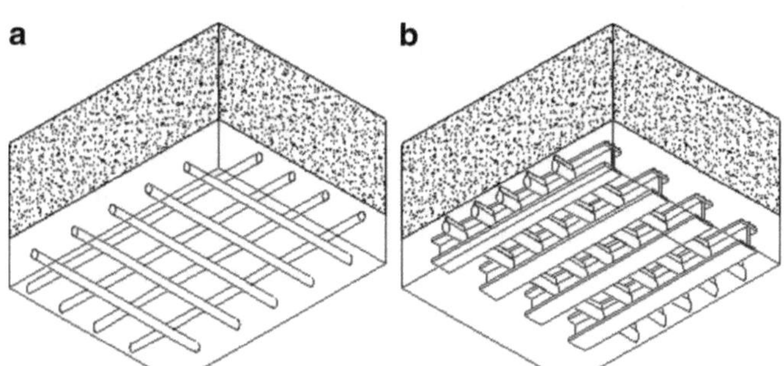

Fig. 8.1 FRP grid (**a**) or grating (**b**) reinforced concrete deck

By expanding this reinforcement concept, three-dimensional FRP grating cages have also been developed for bridge decks. The grating typically consists of an orthogonal grid with longitudinal and transverse glass/polyester or glass/vinylester pultruded bars (as shown in Fig. 8.1b). The longitudinal bars normally have miniature I-shape, T-shape or rectangular cross-sectional profiles and the transverse bars are of circular cross-section. The commercial grating products can be easily found from many grating manufacturers in the nation such as the ones associated with the Fiberglass Grating Manufacturers Council (ACMA 2007).

8.3.1 Structural Performance

Laboratory investigations were conducted during the early stage of the FRP grid applications in bridge slabs (e.g., Bank et al. 1992; Matthys and Taerwe 1995). It was reported that the ultimate loads in the FRP grid-reinforced concrete slabs were equal to or higher than those in the conventional concrete slabs reinforced with steel (Banthia and Al-Asaly 1995). Full-scale test results from two parallel investigations on flat concrete panels reinforced with GFRP and CFRP single grid and bi-grid systems were summarized in a report by Dutta et al. (1998). The report concluded that "although the FRP grid-reinforced concrete is more flexible than steel-reinforced concrete, its pos-failure deformation was pseudo-ductile, characterized by continuous structural deformation through multiple low-level brittle failures before the onset of catastrophic failure". Good strength and serviceability performance was also reported based on the tests using the NEFMAC FRP grids that had a much lighter weight than steel rebars (Matthys and Taerwe 2000; Yost and Schmeckpeper 2001). A bi-layer glass/vinylester FRP pultruded grid system was employed in a series of full-scale concrete deck specimens, which were tested under punching shear and fatigue cycling with a supporting condition simulating the actual field condition (Jacobson et al. 2004). The satisfactory test results demonstrated the feasibility of the concept in the field and the deck system was later constructed in a bridge on US Highway 151 over De Neveu Creek in Wisconsin (Bank et al. 2006).

Quasi-static tests performed on the FRP grating-reinforced concrete system showed satisfactory service load deflection and load carrying capacity (Bank et al. 1992, 1997). The ultimate load carrying capacity of the FRP reinforced slabs exceeded the load at which yielding occurred in the steel-reinforced slab (Bank et al. 1992). Under punching shear, unlike the FRP rod-reinforced slabs, the FRP grating-reinforced concrete slabs did not display a sharp load drop at punching failure. They were still able to absorb energy stably after the initial failure (Bank and Xi 1995). The pultruded GFRP grating sections were also investigated in the medium-scale one-way concrete slab experiments where their behavior was compared with that of concrete slabs reinforced by either steel or FRP bars (Biddah 2006). The test results revealed good composite action between the pultruded fiberglass grating and the concrete. It was concluded that it was feasible to rely on the grating reinforcement in resisting the bending and shear stresses with reasonable ductility in the slab.

Based on these existing research data, the FRP grid- or grating-reinforced concrete decks have shown predictable and reliable structural performance in bridge applications.

8.3.2 Design Concept

The grid and grating design concept itself is not new to astronautical, aeronautical, architectural and civil engineers (Meyer 1973), but its exploitation in bridge construction started to gain success till early 1990s as the FRP composites and efficient manufacturing methods became more available. In an early study conducted by Goodspeed and Aleva (1995), a design procedure of 2-D FRP grid-reinforced concrete slab (with FRP's properties similar to NEFMAC) was outlined following the analysis principles similar to conventional steel-reinforced concrete section where equilibrium and compatibility conditions of the section were satisfied. The analysis was then verified by their laboratory flexural tests performed on flat slabs (with a dimension of 1.22×3.05 mm). For the design of FRP-grid reinforced concrete slabs, Banthia and Al-Asaly evaluated the use of methods based on the various code equations developed for conventional steel-reinforced concrete slabs (Banthia and Al-Asaly 1995) together with some test observations. It was concluded that these equations were applicable to slabs reinforced with FRP grids with only minor modification. Comparable comments were also reached in other studies (Matthys and Taerwe 2000; Yost and Schmeckpeper 2001) where design procedures developed for steel-reinforced concrete slabs were followed except that the design was controlled by the serviceability limit state rather than the ultimate strength limit state.

Although the existing research implies that the FRP grid- or grating-reinforced concrete decks can be designed similarly to the conventional steel-reinforced concrete, special attention should be paid to the following aspects besides the ones previously discussed for FRP rod-reinforced concrete decks:

1. Ductile failure as endorsed in the conventional steel-reinforced concrete design method is more difficult to achieve due to the non-ductile nature of the FRP materials. Therefore, FRP grid- or grating-reinforced concrete decks dissipate less energy, providing less warning in the structure as compared with the steel-reinforced concrete slabs.
2. Due to the orthogonal pattern of the reinforcing bars, the load transfer mechanism of FRP grid- or grating-reinforced concrete slabs is different from the steel-reinforced ones. The internal force transfer now concentrates at the intersecting grid lines rather than being distributed along the rebar line. The grid joints thus need to be properly designed and manufactured to avoid potential larger crack widths at the grid joint locations (Dutta et al. 1998).
3. In cases where splicing of FRP grids and gratings is needed, specialized splicing methods shall be considered rather than simple overlapping (Bank 2006).
4. Device that can prevent the FRP grids or grating from floating during concrete pouring needs to be designed to facilitate the constructability of the system (e.g., PVC inserts with zip-ties as used in Steffen et al. 2001b). Other practical issues

should also be considered during the design stage, such as ensuring a proper grid clearance from concrete surface and vibration of concrete especially in the grid overlapping regions.

8.3.3 Application Status

A wide range of bridge applications on FRP reinforcing grids and gratings have been successfully carried out over the past 10 years in North America. The NEFMAC grid made of GFRP was used to partially reinforce the concrete deck in both the Waterloo Creek Bridge, British Columbia, and the Chatham Bridge, Ontario (Benmokrane and Cousin 2005; ISIS 2010). The field monitoring results showed little deterioration of the GFRP grid in those bridges over the years of their service (Mufti et al. 2005). The Joffre Bridge, located over the St.-François River in Sherbrooke, Quebec, also included the use of CFRP grids in its deck slab and GFRP reinforcing bars in the traffic barrier and sidewalk (Benmokrane et al. 1999). CFRP grids were also used to reinforce the negative bending moment regions of concrete slabs in the Kent County Road No. 10 Bridge (Tadros et al. 1998). The superstructure of the Rollinsford Bridge in Rollinsford, New Hampshire, was made of a cast-in-place deck reinforced with a 2-D NEFMAC rigid CFRP grid (Steffen et al. 2001a, b). The bridge is 33.5 m long, simply supported, and has been in service since December 2000 (also known as the first steel-free concrete bridge made entirely with CFRP grid in the US).

The bridge on the US Highway 151 over the De Neveu Creek in Wisconsin (39.6 m long and 13.7 m wide) was constructed with double-layer prefabricated FRP grids for rapid bridge deck construction in April 2004 (Bank et al. 2006). This double-layer 3-dimensional pultruded FRP grid is 12.9 m long, 2.4 m wide and 16.5 cm deep.

8.4 Continuous Flat or Curved FRP Plate Reinforced Concrete Deck

FRP reinforcement for concrete deck is also available in the form of continuous flat or curved plate, as shown in Fig. 8.2. They are typically integrated to the concrete deck through proper surface treatment and/or the use of additional mechanical

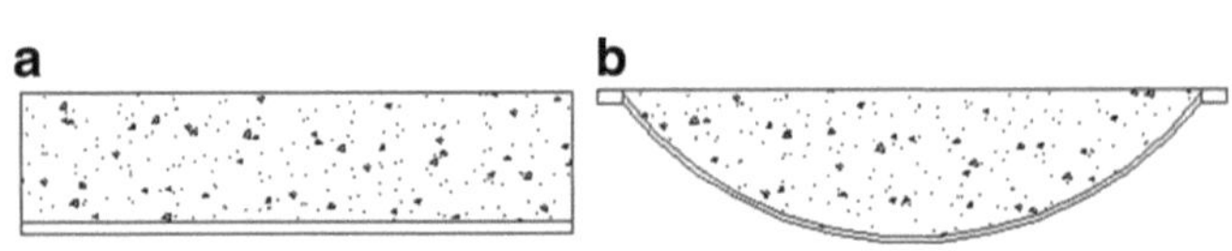

Fig. 8.2 Examples of (**a**) flat or (**b**) curved plate reinforced concrete deck

device. Evaluation of these concrete deck systems are still going through theoretical and laboratory research. Their application status in the field is not yet available.

8.4.1 Structural Performance

The commercially available flat Pultruded GFRP plates were investigated in serving as flexural reinforcement and open form for concrete slabs in replacement of steel (Honickman 2008). The GFRP plate consisted of alternating layers of unidirectional E-glass roving and random mats resulting in a total thickness of 9.5 mm. Polyester resin was used as the adhesive. To improve the interaction between the concrete and GFRP plate, two types of epoxy adhesives with and without silica stones based coarse aggregates, threaded GFRP and steel studs were used in constructing the one-way slab specimens. The test results under flexure showed that failure was mainly due to debonding of the GFRP plate after flexural or shear cracking in the concrete. The wet adhesive bonding of fresh concrete to GFRP plate was the most effective way to achieve desired structural response and fabrication speed (Honickman 2008).

The concept of continuous curved FRP plate was first proposed by Bakeri (1989), where light weight concrete was mixed with hooked-end steel fibers as filling material in glass-fiber reinforced pultruded truss or membrane slabs (i.e., concrete filled membrane deck). The rationale behind this idea was to effectively utilize the tensile strength of FRP and the compressive strength of concrete. The mechanical performance of the system was analytically evaluated using finite element method (Bakeri 1989) and the results showed that a deflection of less than $S/800$ was achieved where S is the stringer spacing upon loading to service load condition (Baekri and Sunder 1990).

8.4.2 Design Concept

Since the FRP reinforcement is in the form of a continuous plate, it can serve as an open structural form for concrete without the need for shoring system during the construction. In cases like this, the flat or curved FRP plate shall be designed to carry the construction load (i.e., self-weight of the fresh concrete and other loads during construction phase). The flat GFRP plate-reinforced concrete slabs as studied in Honickman (2008) were designed following the conventional steel-reinforced concrete members with equivalent strength and stiffness in the slab section. Moreover, to prevent the debonding failure, the shear strength of 1 MPa was recommended for the design value in the adhesive bond to wet concrete in the slab.

The analysis results on the curved FRP plate-reinforced concrete deck (concrete filled membrane deck) showed that the deflection criterion was satisfied but the tensile stress in concrete greatly exceeded the allowable level (Bakeri 1989). The bond strength also needed to be further investigated to further refine this design concept.

8.5 Sandwich Panel with Foam, Balsa or Steel Inserts

The concrete deck design covered in this section refers to the use of sandwich panel reinforcement that is made of FRP sheets and non-structural inserts such as foam, balsa wood, cellular core or steel (see Fig. 8.3 for such examples). These sandwich panels also serve as the formwork for concrete slab during the bridge construction.

8.5.1 Structural Performance

Three-dimensional woven composites are commonly used in aerospace and automobile applications where fiber delamination at joints can generally be avoided. Research led by a group at North Carolina State University investigated the feasibility of fabricating a GFRP sandwich panel using such a three-dimensional woven process and employing the product in constructing concrete bridge decks as flexural reinforcement (Johnson et al. 2007). The GFRP panel consisted of top and bottom skins made of two orthogonal woven fabrics stitched together by fibers in the perpendicular direction to form a 3-D structure. One layer of both skins was cut along the joint and folded 90° to form a vertical web. The cavities enclosed by the webs and skins were filled with four in thick balsa core (Fig. 8.3a). The composite action between the GFRP deck and the concrete topping was achieved through shear connectors installed on the top skin of the deck covering its entire width. The shear connectors were made of similar GFRP woven fabric as the deck and were inserted with wood. Although a steel wire mesh was placed directly on top of the shear connectors in the testing specimens, it was not recommended for panels that will be exposed to the elements during the design life (Johnson et al. 2007). Two deck specimens constructed from this hybrid GFRP/concrete were tested under three-point bending, both of which failed due to failure of the concrete overlay. It was concluded that the performance of the deck was adequate for AASHTO truck load classifications with a limited skin thickness and deck depth. Additional treatment such as surface roughening of the GFRP top skin or use of a bonding agent was recommended to further increase the bond at the GFRP panel-concrete interface.

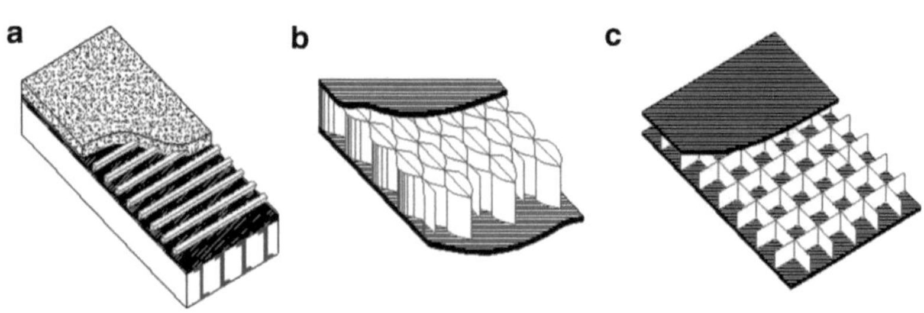

Fig. 8.3 Examples of sandwich panel reinforced concrete deck: (**a**) GFRP/concrete hybrid panel; (**b**) honeycomb core panel; and (**c**) truss (web) core panel

Alagusundaramoorthy et al. (2006) investigated the structural behavior of various FRP composite deck panels including a fiberglass deck panel supplied by Hardcore Composites using the 'cell core' technology in conjunction with SCRIMP. The cell core is foam wrapped with fiberglass fabric to create an internal lattice structure. The composite deck comprised of multiple wrapped cells with upper and lower fiberglass fabric skin faces. It was reported that the tested deck panels satisfied the flexural, shear and deflection performance criteria. Hexagonal honeycomb cores are commonly used in sandwich panel construction where two high density face sheets are adhesively bonded to a low density cellular core structure such as hexagonal honeycomb core (Hohe et al. 2001). Their application in bridge desks has recently been investigated analytically and experimentally (Morcous et al. 2010).

The hybrid concept of FRP-steel core was proposed for composite sandwich panels used in bridge decks in order to not only improve stiffness and buckling response but also be cost efficient compared to all glass GFRP decks (Ji et al. 2009). The composite sandwich deck system was made of wrapped hybrid core of GFRP grid and multiple steel box cells with upper and lower GFRP facings. It was found that the proposed composite sandwich panel with hybrid FRP-steel core was efficient for use in bridges even though the thickness of the hybrid deck could be decreased by 19% when compared with the all GFRP deck. The failure mode of the proposed hybrid deck was found to be more favorable because of the yielding of the steel tube when compared with that of all GFRP decks (Ji et al. 2009).

8.5.2 Design Concept

The idea for using thin but strong face sheets (GFRP laminates) bonded to thicker lightweight core materials (foam, balsa wood, truss or honeycomb cell core, and steel core) allows engineers to build lighter and more durable bridge decks in the field. They can be either used by themselves (polymer deck) or as tensile reinforcement for concrete in the slab. In both cases, the integrity of the deck performance heavily relies on the adhesive bond between the face sheets and the core (Triantafillou and Gibson 1989). Research has also revealed that stress concentration exists at the sheet-core joints due to deformation incompatibility of the face sheets and the core (Hohe et al. 2001). The general analysis and design methods for structural sandwich panels are available (Allen 1969). The honeycomb sandwich deck panels were modeled as a three-layer equivalent plate in the study by Davalos et al. (2001). The good correlation with the test data led to a conclusion that the equivalent orthotropic properties as developed could be used with confidence in design analyses of the FRP sandwich panel (Davalos et al. 2001). Research results obtained by Morcous et al. on bridge decks made of FRP honeycomb sandwich panels showed that simplified I-beam modeling method was efficient and reliable to study the overall performance of the structure (Morcous et al. 2010).

8.5.3 Application Status

The Bennetts Creek Bridge (Route 248) in Steuben County, New York, was constructed in 1998. This single-span bridge is 7.6 m long secured to abutments with steel dowels and consists of hardcore deck from hand lay-up sandwich slab and E-glass face sheet and orthogonal honeycomb core (Alampalli et al. 2002). It is also known as the first state-owned and -operated single-span composite bridge. Other bridge decks systems utilizing structural sandwich panels, to name a few, include the No-Name Creek Bridge in Russell and a bridge deck replacement project in Crawford County, Kansas; the Hanover Bridge deck replacement project and the West Buckeye Bridge in Monongalia, West Virginia; the demonstration project using FRP web core sandwich superstructure by the New York State Department of Transportation (Aref et al. 2005).

8.6 Concrete Decks Reinforced with Stiffened or Corrugated Stay-In-Place Formwork

To further enhance the interaction between the FRP plate and concrete, FRP reinforcing plates are often stiffened with structural sections adhesively bonded to the interface. As illustrated in Fig. 8.4, these structural components can be in a variety of forms (Cheng and Karbhari 2006a): (a) I-beam, (b) vertical web, (c) T-upstand, (d) cavity section, (e) tubular stiffener, (f) bonded tubes and (g) corrugated profile with or without connectors. Besides tensile reinforcement, these continuous FRP-based open forms also provide a natural formwork for the concrete slab to be cast on the top, accelerating the construction speed of the bridge in the field. Therefore, they are often referred to permanent FRP stay-in-place (SIP) formwork, similar to the prefabricated metal SIP profiles that are commonly used for construction of composite floor or deck systems.

8.6.1 Structural Performance

Extensive research has been conducted in evaluating the structural performance of concrete decks reinforced with different types of stiffened or corrugated SIP formworks. A pultruded GFRP SIP form stiffened with small *I-beam ribs* (Fig. 8.4a) was first introduced in 1994 where the form was attached to the bottom of a steel-reinforced concrete slab in order to enhance the durability of the deck system (Ishizaki et al. 1994). Experimental results showed that the deck system had higher durability when comparing to the conventional steel reinforced concrete slabs (Matsui et al. 2001). This SIP plate can also be formed by bonding the bi-directional

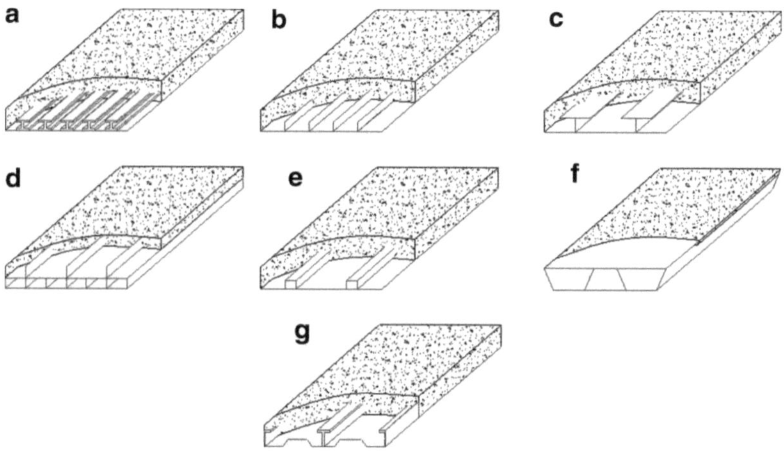

Fig. 8.4 Concrete deck systems reinforced with stiffened or corrugated SIP form: (**a**) I-beam; (**b**) vertical web; (**c**) T-upstand; (**d**) cavity section; (**e**) tubular stiffener; (**f**) bonded tubes; and (**g**) corrugated profile with or without studs

FRP gratings (as discussed in Sect. 8.3) onto a FRP panel, referred to as the gridform system (Ringelstetter et al. 2006). Lab tests showed that this gridform system meets the strength and deflection code requirement. The use of pultruded FRP panel with *T-upstands* (Fig. 8.4c) was proposed by Hall and Mottram in 1998 where the FRP panel section was a commercially available floor panel of non-strandard section. The overall behavior of the system was shown to be linear elastic within the serviceability range.

The *cavity section* of the FRP SIP form (Fig. 8.4d) was utilized in constructing a precast FRP-concrete composite deck system with concrete wedge for enhanced connection between concrete and FRP deck (Park et al. 2009, 2010). Note that longitudinal and transverse steel rebars were still used in this system. Static test results showed an ultimate strength exceeding the design requirement although the final failure was due to the failure of FRP module near the supports. The deflection and crack width criteria were also easily satisfied. This deck system was thus recommended for its use in cable-stayed bridges (Park et al. 2010).

For the SIP form containing *tubular sections* (Fig. 8.4e), three-point bending static test was performed in 1999 on concrete specimens reinforced by this form with different lengths and depths (Harik et al. 1999). The test results indicated that the panels were able to meet the deflection and strength criteria specified by Ohio DOT with the factor of safety (load at failure/load at service) greater than 3 for all panels. The failure of the specimens was found due to the combined flexure and shear and debonding of the GFRP tubular sections from the concrete (Harik et al. 1999). The system was also found to meet the service load design criteria after one million fatigue cycling at a controlled high temperature of 49°C (Lopez-Anido et al. 1999).

The promising test results led more interesting research into this tubular SIP system (Reising et al. 2001; Dieter et al. 2002; Cheng et al. 2005; Cheng and Karbhari 2006b).

The hybrid deck system developed by Kitane et al. (2004) comprises a layer of concrete placed on the compression side of the all GFRP deck section consisting of several trapezoidal GFRP box parts bonded together and wrapped with an FRP laminate to produce an integral section (*bonded tubes*, Fig. 8.4f). With the consideration of reducing initial costs and increasing stiffness of the all GFRP deck, the system was proposed for a one-lane superstructure with a span length of 18.3 m. Static test results on the one-fifth scaled bridge model showed that the model met the stiffness requirement with significant reserve strength and the fatigue testing revealed insignificant stiffness degradation in the system.

A hybrid GFRP/concrete bridge deck consisting of *corrugated pultruded GFRP plate with T-upstands* (studs, as shown in Fig. 8.4g) was investigated recently (Liu et al. 2010). The flexural test results showed that the sand-coating surface treatment on the GFRP plate and the T-upstands improved the connection and the ultimate flexural resistance. Final failure was caused by the debonding of the T-upstands due to large slippage at the interface followed by shear failure initiated from the holes of the T-upstands (Liu et al. 2010).

8.6.2 Design Concept

Based on the existing research, the design of these hybrid FRP-concrete deck systems are primarily driven by the flexural-shear strength of concrete, instead of the stiffness related deflection constraints that are typical for all FRP composite decks. Punching shear and fatigue state do not appear to be the governing limit states for the design of FRP SIP form reinforced concrete deck. Moreover, over-reinforcement of FRP form is desired in order to avoid the sudden brittle type of failure in FRP composites. Preliminary design approaches similar to the one developed for the concrete-CFRP tubular deck system (Cheng and Karbhari 2006c) can be formulated for each system. The general design steps include: (1) design the FRP SIP form to carry concrete and panel weight and the construction live load (static analysis can be conducted using classical beam theory); (2) design the shear-bond strength between the concrete and FRP SIP form; (3) design the flexural strength of the section (based on compatibility, equilibrium, and material constitutive laws); and (4) check deflection and crack width for serviceability criteria.

8.6.3 Application Status

Several of the aforementioned FRP stiffened or corrugated SIP formworks have been adopted in the field applications. For example, the 207 m long five-span Salem Avenue

Bridge on Route 49 in Dayton, Ohio (Reising et al. 2001) contains a 1.5-span hybrid GFRP-concrete deck (the rest of the bridge uses all polymer deck). This hybrid deck was constructed from pouring a concrete slab over pultruded GFRP panels reinforced with GFRP tubular sections and additional GFRP rebars. A two span highway overpass on US Highway 151 was constructed by the Wisconsin DOT in 2003 (Berg et al. 2004). The bridge deck consists of three different forms of GFRP reinforcement: FRP stay-in-place form, deformed FRP bars, and a specially prefabricated FRP reinforcing grid. The deck is supported by five prestressed I-shaped concrete girders.

8.7 Polymeric Decking Systems

Modular bridge decks fabricated totally from polymer matrix composites have been explored since early 1980s. They are available in many commercial and custom formats (as illustrated in Fig. 8.5): Superdeck system via pultrusion (Lopez-Anido et al. 1997), E-Z span system using pultrusion (Williams et al. 2003), Duraspan system by pultrusion (Cassity et al. 2002), Hardcore system via vacuum assistant resin transfer molding (VARTM) process, Kansas system by hand lay-up process, Strongwell system (rectangular hollow sections, Salim and Davalos 1999; Tromp and Souren 2003), fiber-reinforced recycled plastic lumber system using compression molding process, lightweight FRP deck system through pultrusion process, and X-shaped filament wound FRP deck system (X-shaped FRP deck has a configuration formed from assembling the individual diamond and triangular parts that are then bonded on top and bottom with a combination of woven roving and chopped mat, Plecnik and Azar 1991).

Detailed performance data of polymeric bridge deck systems is readily available (Keller 2001; Bank 2006) and therefore is not discussed in this section. Many of these polymeric deck systems have been implemented in field constructions with the use of either conventional steel and concrete or composite girders as primary

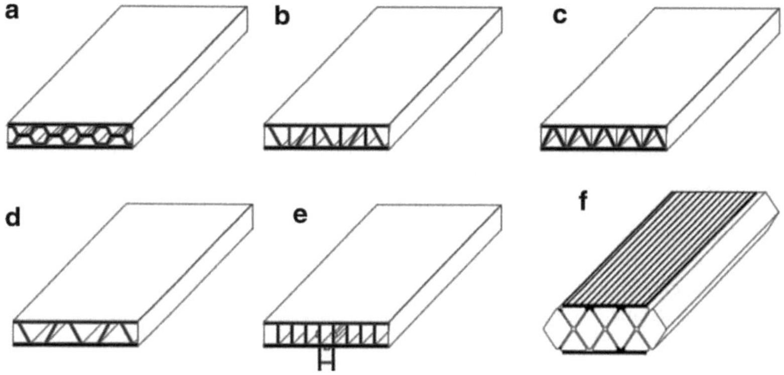

Fig. 8.5 Examples on polymeric decking systems (**a**) Superdeck; (**b**) Duraspan; (**c**) EZspan; (**d**) Fiberline; (**e**) ACCS; and (**f**) X-shaped

supporting members. Noteworthy systems include the bridge system using Superdecks supported by steel and pultruded composite beams (Lopez-Anido et al. 1997); system using multi-cellular deck on steel stringers (Reising et al. 2001); system consisting of 2D and 3D truss deck supported by steel beams (Chiewanichakorn et al. 2003); and system using Strongwell decks placed over wide-flange FRP I-beams (Salim and Davalos 1999). An all-GFRP corrugated-core bridge superstructure was installed and opened to the public in Yeongwol-Gun County, Kangwon-Do, Korea on May 2002 (Ji et al. 2010). More statistical data on most existing polymeric bridges in the world are compiled in the reference provided by the American Composites Manufacturers Association (ACMA 2003).

8.8 Conclusions

The FRP reinforcement offers bridge engineers and practitioners an alternative to solving practical problems of steel-reinforcement in aggressive environments and where magnetic or electrical properties of steel are not desired. The use of FRP materials as primary reinforcement either in pure FRP form or in combination of concrete (steel-free) is reviewed in this chapter. Highlighted are the design concept, structural performance in contrast to conventional steel-reinforced concrete system, and current research and application status of these bridge deck systems. Although promising, the viability of applying FRP reinforcement in concrete bridge decks shall be determined with caution. Design guidelines for each type of deck system still need to be further developed and incorporated into the existing ACI guide which only applies to FRP bar-reinforced concrete structures.

References

AASHTO (2008) Guide specifications for design of FRP pedestrian bridges. American Association of State Highway and Transportation Officials, Washington, DC

ACI (2004) Prestressing concrete structures with FRP tendons. 440.4R-04, American Concrete Institute, Farmington Hills

ACI (2005) Building code requirements for structural concrete and commentary. American Concrete Institute, Farmington Hills

ACI (2006) Guide for the design and construction of concrete reinforced with FRP bars. ACI 440.1R-06, American Concrete Institute, Farmington Hills

ACMA (2007) Fiber Grating Manufacturers Council (FGMC), American Composites Manufacturers Association. http://www.acmanet.org/fgmc/. Cited on 14 Jan 2010

Aiello M, Focacci F, Huang PC, Nanni A (1999) Cracking of concrete cover in FRP reinforced concrete elements under thermal loads. In: Proceedings of the 4th international symposium on FRP for Reinforcement of Concrete Structures (FRPRCS-4), Baltimore, pp. 233–243

Alagusundaramoorthy P, Harik IE, Choo CC (2006) Structural behavior of FRP composite bridge deck panels. J Bridg Eng 11(4):384–393

Alampalli S, O'Connor J, Yannotti AP (2002) Fiber reinforced polymer composites for the superstructure of a short-span rural bridge. Compos Struct 58(1):21–27

Allen HG (1969) Analysis and design of structural sandwich panels. Pergamon Press, Oxford/New York, 283p

American Composites Manufacturers Association (ACMA) (2003) Global FRP use in bridge applications. http://www.mdacomposites.org/mda/bridge_statistics.htm. Cited on 17 Jan 2011

Aref AJ, Almpalli S, He Y (2005) Performance of a fiber reinforced polymer web core skew bridge superstructure. Part I: field testing and finite element simulations. Compos Struct 69(4):491–499

Autocon Composites Inc. (2010) http://www.autoconcomposites.net/NEFMAC.html. Cited on 14 Jan 2010

Bakeri PA (1989) Analysis and design of polymer composite bridge decks. Thesis submitted to the Dept. of Civil Eng. In partial fulfillment of the requirements for the degree of Master of Science, Massachusetts Institute of Technology, Cambridge

Bakeri PA, Sunder SS (1990) Concepts for hybrid FRP bridge deck systems. In: First Materials Engineering Congress. Serviceability and durability of construction materials, Denver, pp 1006–1015

Balendran RV, Rana TM, Maqsood T, Tang WC (2002) Application of FRP bars as reinforcement in civil engineering structures. Struct Surv 20(2):62–72

Bank LC (2006) Composites for construction: structural design with FRP materials. Wiley, Hoboken

Bank LC, Xi Z (1995) Punching shear behavior of pultruded FRP grating reinforced concrete slabs. In: Proceedings, 2nd international symposium on non-metallic (FRP) Reinforcement for Concrete Structures, Ghent, pp 360–367

Bank LC, Zuhan X, Munley E (1992) Tests of full-size pultruded FRP grating reinforced concrete bridge decks. In: Proceedings, materials: performance and prevention of deficiencies and failures; Material Engineering Congress. ASCE, Reston, pp 618–630

Bank LC, Yehoshua F, Shapira A (1997) Three-dimensional fiber-reinforced plastic grating cages for concrete beams: a pilot study. ACI Struct J 94(6):643–652

Bank LC, Puterman M, Katz A (1998) The effect of material degradation on bond properties of FRP reinforcing bars in concrete. ACI Mat J 95(3):232–243

Bank LC, Oliva MG, Russell JS, Jacobson DA, Conachen M, Nelson B, McMonigal D (2006) Double-layer prefabricated FRP grids for rapid bridge deck construction: case study. J Compos Const 10(3):204–212

Banthia N, Al-Asaly M (1995) Behavior of concrete slab slabs reinforced with fiber-reinforced plastic grid. J Mater Civ Eng 7(4):252–257

Benmokrane B, Cousin P (2005) University of Sherbrooke GFRP durability study report. ISIS Canada Research Network, 45 p

Benmokrane B, Masmoudi R, Chekired M, Rahman H, Debbache Z, Tadros G (1999) Design, construction, and monitoring of fiber reinforced polymer reinforced concrete bridge deck. In: The 4th international symposium on fiber reinforced polymer reinforcement for reinforced concrete structures, American Concrete Institute, Detroit, pp 87–102

Benmokrane B, El-Salakawy E, Desgagne G, Lackey T (2004) FRP bars for bridges. Concrete Int 26:84–90

Berg AC, Bank LC, Oliva MG, Russell JS (2004) Construction of a FRP reinforced bridge deck on US highway 151 in Wisconsin. Proc., The 83rd annual transportation research board meeting, CD-ROM, Washington, DC, vol 25, 25 p

Biddah A (2006) Structural reinforcement of bridge decks using pultruded GFRP grating. Compos Struct 74(1):80–88

ISIS Canada Research Network (2010) Intelligent sensing for innovative structures http://www.isiscanada.com/demo/ontario.html. Cited on 16 Jan 2011

Carvelli V, Pisani MA, Poggi C (2010) Fatigue behavior of concrete bridge deck slabs reinforced with GFRP bars. Compos Part B Eng 41(7):560–567

Cassity P, Richards D, Gillespie J (2002) Compositely acting FRP deck and girder system. Struct Eng Int: J Int Assc Bridge Struct Eng (IABSE) 12(2):71–75

Cheng L, Karbhari VM (2006a) Fatigue behavior of a steel-free FRP-concrete modular bridge deck system. J Bridge Eng 11(4):474–488
Cheng L, Karbhari VM (2006b) New bridge systems using FRP composites and concrete: a state-of-the-art review. Prog Struct Eng Mater 8(4):143–154, Wiley, New York
Cheng L, Karbhari VM (2006c) Design approach for a FRP structural formwork based steel-free modular bridge system. Int J Struct Eng Mech 24(5):561–584, Techno-Press
Cheng L, Zhao L, Karbhari VM, Hegemier GA, Seible F (2005) Assessment of a steel-free fiber reinforced polymer-composite modular bridge system. J Struct Eng 131(3):498–506
Chiewanichakorn M, Aref AJ, Alamphalli S (2003) Failure analysis of fiber-reinforced polymer bridge deck system. J Compos Tech Res 25(2):121–129
CSA (2002) Design and construction of building components with fibre-reinforced polymers. CSA-S806-02, Canadian Standards Association, Toronto
Davalos JF, Qiao P, Xu XF, Robinson J, Barth KE (2001) Modeling and characterization of fiber-reinforced plastic honeycomb sandwich panels for highway bridge applications. Compos Struct 52(3–4):441–452
Dieter DA, Dietsche JS, Bank LC, Oliva MG, Russell JS (2002) Concrete bridge decks constructed with fiber-reinforced polymer stay-in-place forms and grid reinforcing. Trans Res Rec 1814, Paper No. 02–3205, pp 219–226
Dutta PK, Bailey DM, Tsai SW, Jensen DW, Hayes JR, McDonald WE, Smart CW, Colwell T, Earl JS, Chen HJ (1998) Composite grids for reinforcement of concrete structures. USACERL Technical Report 98/81, p 169
El-Ragaby A, El-Salakawy E, Benmokrane B (2007) Fatigue life evaluation of concrete bridge deck slabs reinforced with glass FRP composite bars. J Compos Const 11(3):258–268
El-Sayed A, El-Salakawy E, Benmokrane B (2005) Shear strength of one-way concrete slabs reinforced with fiber-reinforced polymer composite bars. J Compos Const 9(2):147–157
Fujisaki T, Nakatsuji T, Sugita M (1993) Research and development of grid shaped FRP reinforcement. In: Nanni A, Dolan CW (eds) Fiber-reinforced-plastic reinforcement for concrete structures – international symposium, ACI SP-138, Vancouver, pp 287–299
Goodspeed C, Aleva G (1995) Stress distribution in FRP grid reinforced members. In: Taerwe L (ed) Non-metallic (FRP) reinforcement for concrete structures – The 2nd international RILEM symposium (FRPRCS-2), Ghent. E&FN Spon, London, pp 298–305
Grace NF (1999) Continuous CFRP prestressed concrete bridges. Concr Int 21(10):42–47
Hall JE, Mottram JT (1998) Combined FRP reinforcement and permanent formwork for concrete members. J Compos Const 2(2):78–86
Harik I, Alagusundaramoorthy P, Siddiqui R, Lopez-Anido R, Morton S, Dutta P, Shahrooz B (1999) Testing of concrete/FRP composite deck panels. In: 5th construction materials congress, ASCE materials engineering division, Cincinnati, pp 351–358
Hassan T, Abdelrahman A, Tadros G, Rizkalla S (2000) Fibre reinforced polymer reinforcing bars for bridge decks. Can J Civ Eng 27(5):839–849
Hohe J, Becker W, Goswami S (2001) Singular stress fields in cellular cores for structural sandwich panels. Compos Struct 53(1):9–19
Honickman HN (2008) Pultruded GFRP section as stay-in-place structural open formwork for concrete slabs and girders. MSc Thesis, Queen's University, Department of Civil Engineering, Kingston, Ontario.
Ishizaki S, Matsui S, Kubo K (1994) A study on FRP permanent form for reinforced concrete slabs. Tech Rep Osaka Uni 44(2196):295–307
Jacobson DA, Bank LC, Oliva MG, Russell JS (2004) Punching shear in fiber-reinforced polymer FRP bi-layer grid-reinforced concrete bridge decks. In: Proceedings of the 83rd annual transportation research board meeting, CD-ROM, Washington, DC
Ji HS, Son BJ, Ma Z (2009) Evaluation of composite sandwich bridge decks with hybrid FRP-steel core. J Bridg Eng 14(1):36–44
Ji HS, Song W, Ma Z (2010) Design, test and field application of a GFRP corrugated-core sandwich bridge. Compos Struct 32(9):2814–2824
Johnson C, Mohamed T, Rizkalla S (2007) Behavior of three-dimensionally woven glass fiber reinforced polymeric bridge deck. Compos Res J 1(2):27–42

JSCE (1997) Recommendation for design and construction of concrete structures using continuous fiber reinforcing materials, Concrete Engineering Series 23. Japan Society of Civil Engineers, Tokyo

Karbhari VM (2007) Durability of composites for civil structural applications. Woodhead Publishing/Maney Publishing, Cambridge

Keller T (2001) Recent all-composite and hybrid fibre-reinforced polymer bridges and buildings. Prog Struct Eng Mater 3(2):132–140

Kitane Y, Aref AJ, Lee GC (2004) Static and fatigue testing of hybrid fiber-reinforced polymer-concrete bridge superstructure. J Compos Const 8(2):182–190

Liu Y, He J, Fan H, Chen A, Dai L (2010) Experimental study on flexural behaviour of hybrid GFRP/concrete bridge deck. In: CICE 2010 – the 5th international conference on FRP composites in civil engineering, Beijing, 27–29 Sept 2010, pp 197–201

Lopez-Anido R, GangaRao HVS, Vedam V, Overby N (1997) Design and evaluation of a modular FRP bridge deck. International composites expo, Composites Institute, Paper 3-E, Nashville, pp 1–6

Lopez-Anido R, Dutta P, Bouzon J, Morton S, Shahrooz B, Harik I (1999) Fatigue evaluation of FRP-concrete bridge deck on steel girders at high temperature. In: The 44th international SAMPE symposium, 23–27 May 1999, pp 1666–1675

MacGregor JG (1997) Reinforced concrete: mechanics and design, 3rd edn. Prentice Hall, Upper Saddle River

Matsui S, Ishizaki S, Kubo K (2001) An experimental study on durability of FRP-RC composite deck slabs of highway bridges. In: Proceedings of the 3rd international conference on concrete under severe conditions: environment & loading, Vancouver, 18–20 June, pp 933–940

Matthys S, Taerwe L (1995) Loading tests on concrete slabs reinforced with FRP grids. In: Taerwe L (ed) Non-metallic (FRP) reinforcement for concrete structures – the 2nd International RILEM Symposium (FRPRCS-2), Ghent. E&FN Spon, London, pp 287–297

Matthys S, Taerwe L (2000) Concrete slabs reinforced with FRP grids: one-way bending. J Compos Const 4(3):145–153

Meyer RR (1973) McDonnell Douglas Astronautics Company, Isogrid Design Handbook, NASA contractor Report, CR-124075, Revision A

Michaluk CR, Rizkalla S, Tadros G, Benmokrane B (1998) Flexural behavior of one-way concrete slabs reinforced by fiber reinforced plastic reinforcement. ACI Struct J 95(3):353–364

Morcous G, Cho Y, El-Safty A, Chen G (2010) Structural behavior of FRP sandwich panels for bridge decks. KSCE J Civ Eng 14(6):879–888

Mufti A, Onofrei M, Benmokrane B, Banthia N, Boulfiza M, Newhook J, Bakht B, Tadros G, Brett P (2005) Durability of GFRP reinforced concrete in field structures. In: Proceedings of the 5th international symposium on FRP for reinforcement of concrete structures (FRPRCS-5), Kansas City

Nanni A (1993) Fiber-reinforced plastic (FRP) for concrete structures: properties and applications. Elsevier, New York

Nanni A (2001) Relevant field applications of FRP composites in concrete structures. In: Proceedings of the international conference composites in construction – CCC2001, Portugal, pp 661–670

Park SY, Cho Y, Kim ST, Cho JR, Kim BS (2009) Structural behavior of FRP-concrete composite deck with concrete wedge. In: The 9th international symposium on Fiber Reinforced Polymer Reinforced for Concrete Structures (FRPRCS-9), Sydney, 13–15 July 2009

Park SY, Cho K, Kim ST, Cho JR, Kim BS (2010) Structural performance evaluation of precast FRP-concrete composite deck with concrete wedge for cable-stayed bridge. In: CICE 2010 – The 5th international conference on FRP composites in civil Engineering, Beijing, 27–29 Sept 2010, pp 211–214

Plecnik JM, Azar WA (1991) Structural components, highway bridge deck applications. Int Ency Compos 6:430–445

Reising RMW et al (2001) Performance of five-span steel bridge with fiber-reinforced polymer composite deck panels. Trans Res Rec 1770, Paper No. 01–0337, pp 113–123

Ringelstetter TE, Bank LC, Oliva MG, Russell JS, Matta F, Nanni A (2006) Development of a cost-effective structural FRP stay-in-place formwork system for accelerated and durable bridge deck construction. Design of Structures, Transportation Research Record, No. 1976, Transportation Research Board, pp 183–189
Rizkalla SH (1997) A new generation of civil engineering structures and bridges. In: Proceedings of the 3rd international symposium on non-metallic (FRP) Reinforcement for Concrete Structures (FRPRCS-3), vol 1, Japan Concrete Institute, Tokyo, pp 113–128
Salim HA, Davalos JF (1999) FRP composite short-span bridges: analysis, design and testing. J Adv Mater 31(1):18–26
Sen R, Shahawy M, Sukumar S, Rosas J (1998) Effect of tidal exposure on bond of CFRP rods. In: Saadatmanesh H, Eshani MR (eds) Second international conference on composites in infrastructure, vol II, University of Arizona, Tucson, pp 512–523
Steffen RE, Trunfio JP, Bowman MM (2001a) Performance of a bridge deck reinforced with CFRP grids in Rollinsford, New Hampshire, USA. In: Figueiras J, Juvandes L, Furia R (eds) Proceedings of CCC, FRP Composites in Construction, Porto, pp 671–676
Steffen R, Scott D, Goodspeed C, Bowman M, Trunfio J (2001b) Design issues and constructability of a CFRP grid reinforced bridge deck. In: Azizinamini A, Yakel A, Abdelrahman M (eds) High performance materials in bridges. ASCE, Reston, pp 106–116
Tadros G, Tromposch E, Mufti A (1998) University drive/Crowchild Trail Bridge superstructure replacement. In: Proceedings of the 2nd international conference on composites in infrastructure (ICCI-98), vol 1, Tucson, pp 693–704
Triantafillou TC, Gibson LJ (1989) Debonding in foam-core sandwich panels. Mater Struct 22:64–69
Tromp E, Souren WHM (2003) Design of a composite draw bridge. In: Proceedings of ICCM-14, San Diego, pp 1–9
Vijay PV, GangaRao HVS (2001) Bending behavior and deformability of glass fiber-reinforced polymer reinforced concrete members. ACI Struct J 98(6):834–842
Williams B, Shehata E, Rizkalla SH (2003) Filament-wound glass fiber reinforced polymer bridge deck modules. J Compos Const 7(3):266–273
Yost JR, Schmeckpeper ER (2001) Strength and serviceability of FRP grid reinforced bridge deck. J Bridg Eng 6(6):605–612

Chapter 9
Load Testing and Load Distribution of Fiber Reinforced, Polymer Strengthened Bridges: Multi-year, Post Construction/Post Retrofit Performance Evaluation

John J. Myers, David Holdener, and Wesley Merkle

Abstract Fiber-Reinforced Polymer (FRP) strengthening of structural systems has increased in applications over recent years. The many benefits that the technology offers are significant. Field validation of FRP-strengthened bridges through load testing provides a means of measuring the performance of a bridge over time. Non-contact optical surveying equipment is one such method that can measure the deflection of bridges subjected to a static load. Since 1999, more than 25 FRP bridges in Missouri have either been repaired or constructed using FRP products to serve as demonstration and case study projects. A remaining question related to this technology is their long-term field performance. This work presents the load testing results of four bridges located in Morgan County, Crawford County, Iron County, and Dallas County, Missouri. These bridges were originally strengthened in 2003 with different FRP technologies and subsequently load tested biennially. The investigation focused primarily on determining if the bridges had undergone any degradation in the FRP material properties based upon the structures' response to loading. Deflection and load distribution between girders was monitored. Pre- and post-load testing results were compared to better understand the performance over time and study structural degradation. Finally, the visual inspection results of these bridges by the researchers are presented.

FRP is slowly making strides within the Civil Engineering community. The widespread acceptance of these materials is hampered by their unproven long-term

J.J. Myers (✉) • D. Holdener • W. Merkle
Department of Civil, Architectural, and Environmental Engineering,
Missouri University of Science and Technology, 325 Butler Carlton Hall,
1401 North Pine Street, Rolla, MO 65409, USA
e-mail: jmyers@mst.edu; holdener@umr.edu; wmerkle@structurepoint.com

R. Jain and L. Lee (eds.), *Fiber Reinforced Polymer (FRP) Composites for Infrastructure Applications*, Strategies for Sustainability, DOI 10.1007/978-94-007-2357-3_9,

reliability. Monitoring deflections of bridges constructed with FRP technologies under static load testing can help validate the long term performance of these materials.

Keywords Fiber reinforced polymers (FRP) for bridge repair • Load distribution • Load testing and monitoring • Load testing sensitivity • Visual bridge inspection

9.1 Introduction

The transportation infrastructure in the United States is often taken for granted by most citizens; it can be considered the backbone of America because without it, our nation's economy, as well as our lifestyle, would simply cease to exist. Bridges are considered to be a vital part of the infrastructure system. As many structures reach a critical age and become deficient or obsolete, they must be repaired or replaced (Koenigsfeld and Myers 2003). Over 40% of the nation's bridges are estimated to be in need of repair or replacement (Stone et al. 2002). The Federal Highway Administration (FHWA) claims that 25% of our nation's bridges are either deficient or functionally obsolete; in the State of Missouri, this number jumps to 35% (USDOT 2003). Many states, including Missouri, cannot afford to tend to all the repair and replacement needs immediately, so the bridge owners post load restrictions as a temporary solution to the problem.

Bridges become deficient or obsolete for a variety of reasons. Bridge decks typically require major repair or replacement every 15–20 years, while the substructure and superstructure tend to last 40 years or more (Koenigsfeld and Myers 2003). Deicing salt applications cause sapling of concrete, while water seeping through cracks in the concrete corrode unprotected reinforcing steel. Over the past decades, standard truck design weights and traffic volumes have increased; as a result, older bridges that are in good condition do not have the load carrying capacity required for today's traffic needs. Safety considerations have changed as well, and most roadways today are designed to have 12-foot wide lanes. Both changing traffic demands and an already decaying infrastructure require engineers to find alternative solutions on a slim budget.

Fiber Reinforced Polymer (FRP) composites have enjoyed growing popularity over the past decade in repair and strengthening applications of existing bridges and buildings. Use of FRP technologies for such applications was well known for benefits including low cost, fast installation, minimal traffic disruption, and anticipated long-term durability. With the benefits of such a temporary strengthening system in mind, engineers are studying methods to encourage use of these technologies on real structures.

In the case of bridge rehabilitation with new FRP strengthening technologies, diagnostic load testing is required to verify the effects of remediation on the bridge (ACI 437R-03 2002). Load testing is an effective means of evaluating the performance of a structure (AASHTO 2002). Protocol for load testing should be carefully planned; however, no guidelines for load testing currently exist, so testing procedures are determined by the entity performing the test (AASHTO 2002).

9.1.1 Load Testing Overview

Load testing is observing and measuring the response of a structure subjected to controlled loads in the elastic range (AASHTO 2003). Load testing has many purposes and benefits to match; it allows engineers to determine unknown factors such as composite action and unintended fixity levels, continuity, and participation of other members. Load testing is proven to be an effective way for structural engineers to evaluate the performance of a bridge (AASHTO 2002). Several researchers have investigated full-scale load testing of bridges (Vurpillot et al. 1998; Benmokrane et al. 1999; Yang and Myers 2003).

Load testing has two main alternatives: (1) diagnostic testing and (2) proof testing. A diagnostic load test is chosen to find the response characteristics of a structure, including load response, load distribution, validating analytical models, and evaluating the effectiveness of structural repairs and upgrades where the structure may not be accurately load rated (AASHTO 2003). This project utilizes diagnostic load testing to validate the effectiveness of FRP strengthening application. Alternatively, the proof load test is designed to establish the safe load carrying capacity of the structure while loading the structure within the elastic range (AASHTO 2003). Alternatively, the load rating for a bridge can be determined via a proof load test.

Diagnostic and proof load tests can be performed using static or dynamic loads, or a combination of both. Static loads are stationary loads; they induce no vibrations. The location and magnitude of the loads may change during testing. Using standard loaded trucks to park on the structure tends to be easiest for testing existing structures. Dynamic load tests use moving loads to induce vibrations and find a very rapid loading response of the structure. These tests allow for the analysis of dynamic load allowance, frequency, vibration, and a stress range for fatigue evaluation (AASHTO 2003). Dynamic loading is most easily accomplished by loading trucks and driving them over pre-determined locations at given speeds. The following section describes the five bridges that were strengthened with various FRP systems and the load testing techniques.

9.1.2 Bridge Characteristics, Strengthening, and Monitoring Descriptions

9.1.2.1 Bridge Descriptions

The following sub-sections briefly describe the five Missouri Department of Transportation (MoDOT) Bridges that were strengthened with FRP and subsequently load tested. Table 9.1 provides a summary of the strengthening type and location for the respective bridges. Additional details about the bridges and specifics on the design and strengthening systems used may be found in two detailed reports: "Preservation of Missouri Transportation Infrastructure: Validation of FRP Composite

Table 9.1 FRP flexural and shear strengthening types and location with analytical capacity improvement

Bridge id	Span #	Girder	Flexural reinforcing description (girder)	Capacity increase
X-596[g]	2	Interior	ML: 4 plies 20″ wide; NSM bars: 4 total	42%
X-596[g]	2	Exterior	None	NA
X-596[g]	1, 3	Interior	ML: 4 plies 20″ wide; NSM bars: 4 total	44%
X-596[g]	1, 3	Exterior	ML: 2 plies 16″ wide	16%
T-530[g]	1, 3, 5	Interior	ML: 4 plies 16″ wide	29%
T-530[g]	1, 3, 5	Exterior	ML: 2 plies 16″ wide	15%
T-530[g]	2, 4	Interior	1 laminate plate: 12″ wide	29%
T-530[g]	2, 4	Exterior	1 laminate plate: 12″ wide	15%
X-495[g]	2	Interior	ML: 5 plies 20″ wide	40%
X-495[g]	2	Exterior	None	NA
X-495[g]	1, 3	Interior	ML: 5 plies 16″ wide; NSM bars: 2 total	44%
X-495[g]	1, 3	Exterior	ML: 2 plies 16″ wide	16%
P-962[g]	1, 2	Interior	ML: 5 plies 16″ wide plus 4 NSM bars	56%
P-962[g]	1, 2	Exterior	ML: 3 plies 16″ wide	25%
P-962[g]	3	Interior	SRP 3X2: 3 plies 16″ wide	54%
P-962[g]	3	Exterior	SRP 3X2: 2 plies 16″ wide	49%
			Flexural reinforcing description (slab)	
X-596[s]	2	NSM tape	2 per groove at 12″ O/C	78%
X-596[s]	1, 3	Manual layup	1 ply 6″ wide at 15″ O/C	61%
T-530[s]	1, 3, 5	Manual layup	1 ply 9″ wide at 15″ O/C	141%
T-530[s]	2, 4	Laminate plates	1 plate 3″ wide at 15″ O/C	143%
X-495[s]	1, 2, 3	Manual layup	1 ply 6″ wide at 14″ O/C	65%
Y-298[s]	1, 2	Manual layup	2 plies 8″ wide at 12″ O/C	23%
P-962[s]	3	SRP 3X2	1 ply 4″ wide at 20″ O/C	62%
P-962[s]	1, 2	Manual layup	1 ply 6″ wide at 14″ O/C	64%
			Shear reinforcing description (girder)	
X-596[g]	2	Interior	ML: 1 ply continuous U-wrap	26%
X-596[g]	2	Exterior	None	NA
X-596[g]	1, 3	Interior	ML: 2 ply continuous U-wrap	52%
X-596[g]	1, 3	Exterior	None	NA
X-495[g]	2	Interior	ML: 1 ply continuous U-wrap	30%
X-495[g]	2	Exterior	None	N/A
X-495[g]	1, 3	Interior	ML: 2 plies continuous U-wrap	51%
X-495[g]	1, 3	Exterior	ML: 1 ply 12″ wide U-wrap at 24″ O/C	18%
P-962[g]	1, 2	Interior	ML: 4 plies continuous U-wrap	64%
P-962[g]	1, 2	Exterior	ML: 1 ply continuous U-wrap	24%
P-962[g]	3	Interior	SRP 3SX: 3 plies continuous U-wrap	63%
P-962[g]	3	Exterior	SRP 3SX: 1 ply continuous U-wrap	36%

Key: *s* slab/deck strengthening, *g* girder strengthening, *ML* manual layup, *NSM* near surface mounted bar, *SRP* steel reinforced polymer

Fig. 9.1 Tested span of bridge X-596

Technology through Field Testing – Volume I" (2004) and a report by Merkle and Myers (2004a). Only a brief overview of the bridges is presented here as an introduction to the load testing and monitoring of the bridges. A structural analysis was performed using HS20-44 and 3S2 truck loading cases following AASHTO specifications for each bridge to determine the appropriate retrofit requirement (see Table 9.1). Strengthening was designed following ACI 440.2R-02 (2002) guidelines for each of the bridges.

Bridge X-596

Bridge X-596 carried Route C over Lander's Fork Creek in Morgan County, Missouri (MoDOT District 5). The bridge average daily traffic (ADT) was 2000; it was load-posted for trucks over 18 tons to travel 15 miles per hour over the bridge. The bridge was built in 1946 as a Deck Girder/Reinforced Concrete Tee Beam Structure. The bridge had three spans: 42.5, 52.5, and 42.5 ft, totaling 137.5 ft with no skew angle. All spans were simply supported three-span deck girders, or three Tee Beams spaced 9 ft on center, with diaphragms at midspan. The deck slab was monolithically cast 6 in deep and 23.6 ft wide; the roadway was 20 ft wide and striped as a one- lane bridge (see Fig. 9.1). The bridge's visual inspection rating was as follows: deck 6, superstructure 5, and substructure 5. Concrete cracking, exposed reinforcement, and concrete deterioration were found in various locations throughout the deck, superstructure, and substructure. Rusted steel bearing plates were found on the end bents. Concrete and steel samples were taken from the bridges and tested. The concrete compressive strength was found to be 6,000 psi, and the yield strength of the reinforcing steel was found to be 40 ksi. The girders were strengthened for shear as needed. The bents were also strengthened for shear and flexure.

Fig. 9.2 Tested span of bridge T-530

Bridge T-530

Bridge T-530 carried Route M over Crooked Creek in Crawford County, Missouri (MoDOT District 9). The bridge ADT was 200; it was load posted for trucks over 21 tons to travel 15 miles per hour over the bridge. The bridge was built in 1939 as a Deck Girder/Reinforced Concrete Tee Beam Structure. The bridge had five spans each 47.5 ft long, totaling 237.5 ft on a 30° skew. All spans were simply supported four Tee Beams spaced 6.5 ft on center, with diaphragms at mid-span. The deck slab was monolithically cast 6 in deep and 23 ft wide; the roadway was 20 ft wide and striped as a one-lane bridge (see Fig. 9.2). The bridge's visual inspection rating was as follows: deck 5, superstructure 5, and substructure 5. Some concrete deterioration and minor concrete cracking existed throughout the structure; the bent caps had cracking from steel corrosion. The concrete compressive strength was found to be 6,250 psi, and the yield strength of the reinforcing steel was found to be 40 ksi. Flexural strengthening schemes were implemented for the deck and girders. The girders were not strengthened for shear; however, FRP U-Wraps were installed to help anchor the flexural reinforcement. The bents were not strengthened.

Bridge X-495

Bridge X-495 carried Route C over Crane Pond Creek in Iron County, Missouri (MoDOT District 9). The bridge average daily traffic (ADT) was 300; it was load-posted for trucks over 19 tons to travel 15 miles per hour over the bridge. The bridge was built in 1948 as a Deck Girder/Reinforced Concrete Tee Beam Structure. The bridge had three spans: 42.5 ft, 52.5 ft, and 42.5 ft totaling 137.5 ft with a 30° skew. All spans were simply

Fig. 9.3 Tested span of bridge X-495

supported three-span deck girders, or three Tee Beams spaced 9 ft on center, with diaphragms at mid-span. The deck slab was monolithically cast 6 in deep and 23 ft, 7 in wide; the roadway was 20 ft wide and striped as a one-lane bridge (see Fig. 9.3). The bridge's visual inspection rating was as follows: deck 6, superstructure 6, and substructure 7. The concrete and all of the bridge's main structural components were found to be in good condition. The concrete compressive strength was found to be 5,450 psi, and the yield strength of the reinforcing steel was found to be 40 ksi. Flexural strengthening schemes were implemented for the deck and girders. The girders were strengthened for shear as needed. The bents were also strengthened for shear and flexure.

Bridge P-962

Bridge P-962 carried Route B over Dousinbury Creek in Dallas County, Missouri (MoDOT District 8). The bridge ADT was 350; it was load posted for trucks over 18 tons to travel 15 miles per hour over the bridge. The bridge was built in 1956 as a Deck Girder/Reinforced Concrete Tee Beam Structure. The bridge had three spans each 42.5 ft long, totaling 127.5 ft on a 15° skew. All spans were simply supported four Tee Beams spaced 9 ft on center, with diaphragms at mid-span. The deck slab was monolithically cast 6 in deep and 23.7 ft wide; the roadway was 20 ft wide and striped as a two-lane bridge (see Fig. 9.4). The bridge's visual inspection rating was as follows: deck 7, superstructure 6, and substructure 6. The concrete and all of the bridge's main structural components were found to be in good condition. The concrete compressive strength was found to be 6,850 psi, and the yield strength of the reinforcing steel was found to be 40 ksi. Flexural strengthening schemes were implemented for the deck and girders. The girders were strengthened for shear as needed. The bents were also strengthened for shear and flexure.

Fig. 9.4 Side view of bridge P-962

Fig. 9.5 Side view of bridge Y-298

Bridge Y-298

Bridge Y-298 carried Route U over Crews Branch Creek in Pulaski County, Missouri (MoDOT District 9). The bridge ADT was 1,100; it was load posted for trucks over 18 tons to travel 15 miles per hour over the bridge. The bridge was built in 1937 as a Reinforced Concrete Slab Structure. The bridge had two continuous spans each 15 ft long, totaling 30 ft on a 45°skew. The deck slab was monolithically cast 7 in deep and 27.2 ft wide; the roadway was 24 ft wide and striped as a two-lane roadway (see Fig. 9.5). The bridge's visual inspection rating was as follows: deck 5, superstructure 5, and substructure 5. The concrete in the east span was in poor condition, with evidence of water seepage and some permanent deformation. The concrete in the west span was in more sound condition. Longitudinal cracking and exposed reinforcement was found in various locations. The concrete compressive strength was assumed to be 4,000 psi, and the yield strength of the reinforcing steel was assumed to be 40 ksi. Flexural strengthening schemes were implemented for the deck. The strengthening was oriented in the same direction as the existing reinforcing steel. The supports were not strengthened.

9.1.3 Instrumentation Program

The principal objective of the current research project was to retrofit five structurally deficient Missouri bridges with FRP composites as the primary strengthening system. Load testing was to be carried out in an effort to verify that the FRP strengthening system was performing as expected. All five bridges were scheduled for load testing biennially for 5 years following strengthening applications. Because of existing site conditions at four of the five bridges, traditional serviceability monitoring techniques, like using LVDT systems, would be extremely difficult; the search for a better system was initiated.

A variety of high-precision optical, laser methods and other surveying tools are commonly used to measure deflections on bridges (AASHTO 2003). Traditional surveying equipment, similar to older generations of total stations, were tried in the past to measure deflection, but the equipment was ultimately not precise enough to measure deflection on stiff structures with short spans (Yang and Myers 2003). In the last few years, the field of surveying has made significant technological advances. These advances in technology have made it possible to use surveying equipment, like the total station and GPS, to monitor deformation in structural monitoring applications (Roberts et al. 2004). The Leica TCA 2003® total station now measures with much more accuracy than its predecessors (Leica 2004). Robotics further increase the accuracy of this instrument. This model is commonly used in absolute monitoring applications (Leica 2004).

Traditional land surveying techniques utilized many components, with the total station as the principal device. The total station (see Fig. 9.6) was set atop a secure tripod in a location with an unobstructed view of the field targets. Reference points (see Figs. 9.7–9.9) or backsites were set in place to transfer a horizontal angle or an elevation from the reference point to the total station, and then from the total station to the targets (see Figs. 9.8 and 9.9). The reference points also served to check that the total station had not moved. The reference point was a prism mounted atop a secure tripod; the targets were prisms fixed atop a metal rod (range pole). In the case of load testing, the prisms were fixed to points on the tested structure. Prior to implementing the use of this system several case studies were investigated in the laboratory and in-situ to compare its performance to other traditional systems. The performance of the total station in these preliminary validation studies was found to be acceptable (Merkle and Myers 2004b). After several comparative evaluations, the accuracy is generally better than 0.005 in. The major drawback is its limitation to monitor dynamic load tests and the time required to read points and acquire data.

Fig. 9.6 Leica TCA 2003

Fig. 9.7 Reference point

9.1.4 Load Testing Program

The monitoring plan called for load testing the critical locations of each individual structure. Unless the bridge had identical span lengths or utilized varied strengthening systems on a particular span, the longest span was selected for load testing as it would yield the highest stress and deflection values. In most cases, the total station was the only viable system to measure deflection due to girder height and or water restrictions below the structure. Table 9.2 summarizes the bridge designation with

Fig. 9.8 Target (prism) on a bridge

Fig. 9.9 Total station test setup

the respective span length that was load tested. All bridges were load tested prior to strengthening and biennially thereafter for a contract period of 5 years. Each bridge was load tested using two H20 configuration dump trucks provided by MoDOT. The trucks were loaded up to 60 kips each, with most of the weight carried by the rear axles. The trucks were parked in five separate locations on the selected span; the locations were determined to maximize stresses in the structure.

Stop 1 produced maximum shear loading condition, where the trucks were parked side-by-side with their rear axles near one end of the supports. The distance from the support was chosen based on placing the back-most wheels a minimum distance equal to the depth of the section. This was assumed to ensure that the trucks' weight

Table 9.2 Bridge designation and span length load testing information

Bridge id	Load testing span (total # spans)	Span length (feet) Span #1	Span #2	Span #3	Span #4	Span #5
X-596[a]	Span #2 (3)	42.5	**52.5**	42.5	NA	NA
T-530[a]	Span #3 (5)	47.5	47.5	**47.5**	47.5	47.5
X-495[a]	Span #2 (3)	42.5	**52.5**	42.5	NA	NA
Y-298[a]	Spans #1 & 2 (2)	**15**	**15**	NA	NA	NA
P-962[a]	Span #2 (3)	42.5	**42.5**	42.5	NA	NA

NA Not Applicable, conversion factor: 1 m=3.281 ft, [a]simply supported bridge

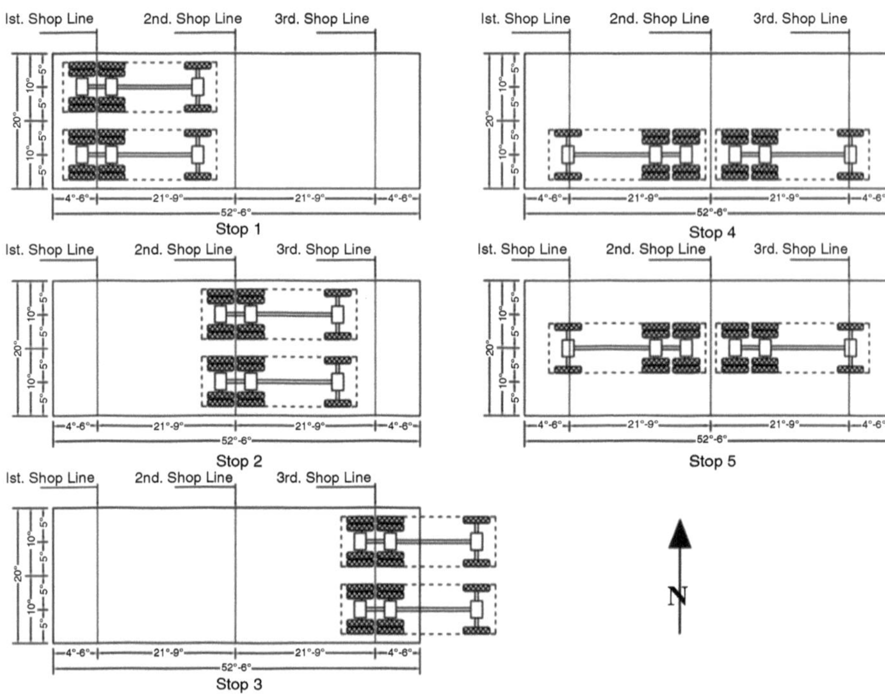

Fig. 9.10 Bridge X-596 truck stops

would develop shearing stresses in the span instead of only loading the support underneath. Stop 2 produced maximum moment loading condition, where the trucks were parked side-by-side with their rear axles centered at mid-span. Stop 3 produced maximum shear loading at the opposite end of the span condition, similar to Stop 1. Stop 4 produced an overload condition on the fully monitored exterior girder, where the trucks were parked back-to-back closest to the barrier wall. Stop 5 produced a similar overload condition on the interior girder, where the trucks were parked back-to-back with their rear axles centered at the bridge's centerline. Figure 9.10 shows the procedure for Bridge X-596 as a representative example. The results from the high moment Load Stops 2, 4, and 5 are reported below.

9.2 Data Processing

Adjustments in readings were undertaken to account for thermal effects and normalization of truck weights for comparison. In order to appropriately account for the camber of the bridge due to thermal effects during testing, an appropriate baseline was established for each stop during testing. The thermal effects were accounted for by taking a weighted average of the first no-load stop and the final no-load stop based upon when the load test stop occurred. On a sunny day, for example, throughout the duration of load testing, camber growth occurred due to thermal expansion. Figures 9.11 and 9.12 illustrate the representative camber growth from start to end of a representative load test and the adjustment for a given load test based on the reading time of the load test stop.

To appropriately compare load tests, truck weights were normalized using the ratio of the theoretical deflection of the first test to the theoretical deflection of the test after strengthening and multiplying this ratio by the experimental deflection recorded during the test after strengthening. The normalized deflection is shown in the following equation:

$$Normalized\,\Delta_{\text{Test}\,i} = \left(\frac{Theoreteical\,\Delta_{\text{Test}\,1}}{Theoreteical\,\Delta_{\text{Test}\,i}}\right) Experimental\,\Delta_{\text{Test}\,i} \tag{9.1}$$

The ratio of theoretical girder deflections was determined by neglecting shear deformations and assuming an elastic, uncracked section subjected to point loads from the truck wheel loads. The transverse distribution factors for each girder were determined by assuming a truck wheel load of 0.5 at each truck wheel location. Additional details regarding this process may be referenced in the report by Myers et al. (2008).

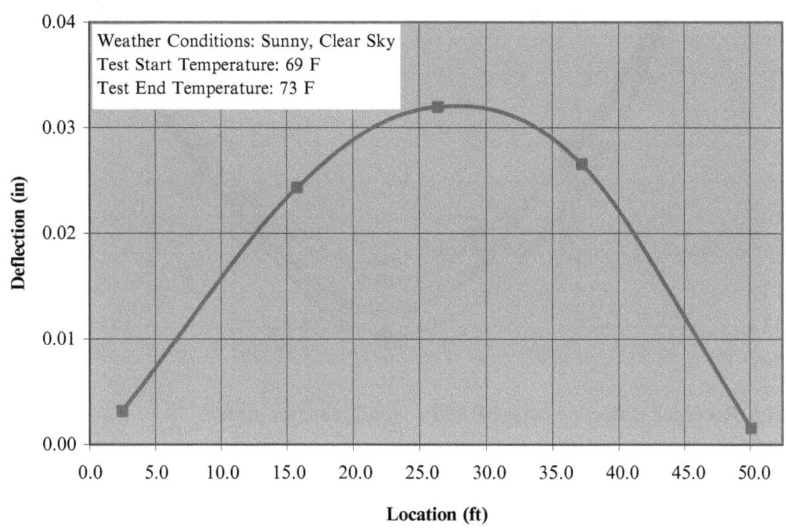

Fig. 9.11 Bridge X-596 thermal camber

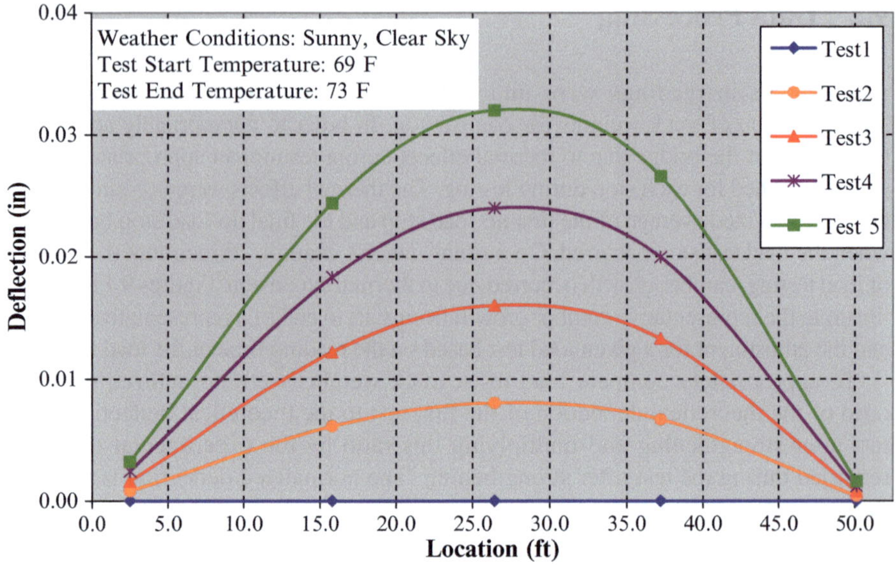

Fig. 9.12 Bridge X-596 adjustment of baseline due to thermal effects

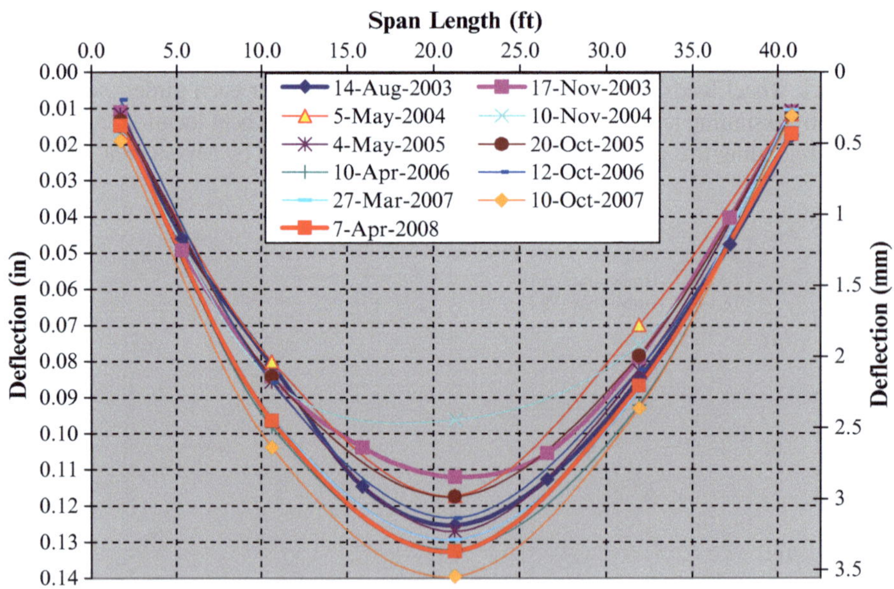

Fig. 9.13 Recorded deflection bridge P-962 – stop 2, exterior girder

Once the ratio of the theoretical deflection was determined for each prism location for all tests, multiplying the ratio by the deflection obtained during testing normalized the deflection values. The normalized deflection values from the post-strengthening tests were then compared to Test 1. Figure 9.13 shows the recorded

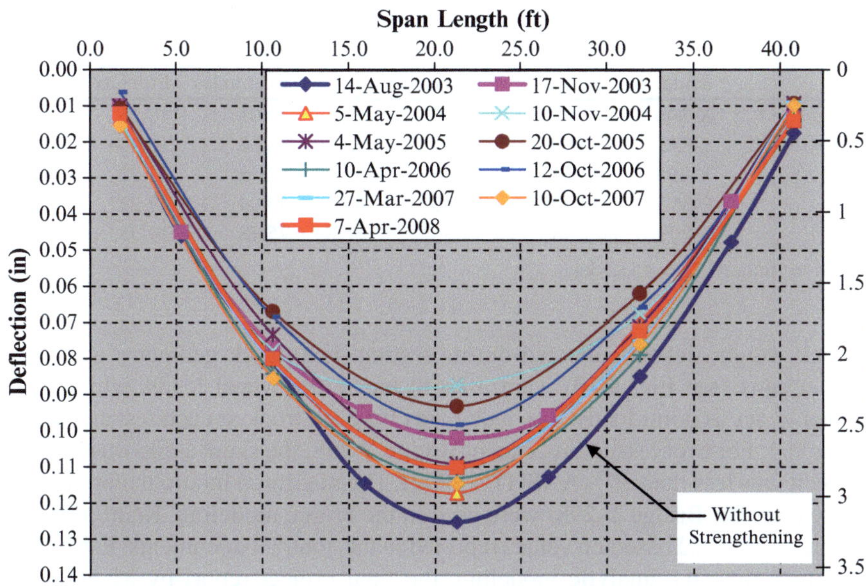

Fig. 9.14 Normalized deflection bridge P-962 – stop 2, exterior girder

deflection data before normalization from Stop 2, the exterior girder of the bridge, for all 11 tests conducted on bridge P-962. Figure 9.14 shows the data after normalization. Notice that Test 1 (before strengthened test) shows an apparent increase in deflection after normalization. This is due to increased truck weights after the first test was conducted. Normalizing the data as illustrated helped to account for the variance in truck weights based upon the provided front and rear tandem axle weights. The axle weights were divided evenly between wheels on each axle; thus there was no way to account for differences in individual wheel loads which could be induced by a load shifted to one side. Additionally, the truck axle loads imposed upon the bridge could vary from the actual recorded weight due to the truck's load shifting during transportation.

9.3 Analytical Modeling

Three (3) Mass-section Tee-Beam models and two (2) Finite Element Modeling (FEM) were created to accurately predict the deflections produced from load testing. The Mass-section Tee-Beam models examined three cases of barrier section engagement (full, half, and none). Similarly, the FEM models examined the bridges with and without engagement of the barrier. The models, without the barrier engagement, ignore composite action of the barrier, which represents the case where longitudinal cracks developed overtime. The models used basic calculations to find pre-strengthened deflection measurements. The loads used were taken from

Table 9.3 Representative truck axle loads (load test 1)

		Truck 1 (kips)			Truck 2 (kips)		
	Bridge id	Rear axles	Front axle	Total	Rear axles	Front axle	Total
Before strengthening	X-596	41.03	17.59	58.62	40.01	17.15	57.16
	T-530	40.90	16.80	57.70	38.32	16.40	54.72
	X-495	42.32	12.76	55.08	42.48	12.72	55.20
	P-962	38.72	13.36	52.08	43.04	17.22	60.26
	Y-298	25.48	11.35	36.83	25.48	11.35	36.83

Conversions units: 1 kip=453.6 kg

the truck weight tickets collected during testing. H20 truck geometry and truck stop diagrams (see Fig. 9.10) were used to locate the wheel loads, which were assumed to act as point loads. The axle loads for each truck at each test are shown in Table 9.3. For every structure, except Bridge Y-298, the skew angle was ignored because it was less than 30° (AASHTO 2002). The measured in-situ material properties (except for Bridge Y-298) were used in the bridge modeling. Reinforcement quantities were used based on values reported in the contract documents. For further details regarding the analytic modeling, the final project report by Myers et al. (2008) may be referenced.

9.4 Analytical Models and Experimental Results

This section compares the theoretical models defined in the previous section with the measurements obtained in the field. Only pre-strengthening testing was used as the models were developed considering the original un-strengthened structure. The results are discussed and compared with visual inspection ratings of each structure.

9.4.1 Comparison Methods

For comparison, only the mid-span deflections of the interior and exterior girders (monitored with nine targets per girder) were used. The calculated deflections for each bridge and each loading scenario were taken from all three models discussed in the previous section. Only the maximum flexural testing (truck stops 2, 4, and 5) data was used because the deflection for maximum shear loading was minimal. Shear contribution to deflection was also assumed insignificant in comparison to the flexural contribution. For each theoretical value, a percent error was computed; this was the difference between theoretical and experimental values divided by the experimental value. A positive error value meant that the theory over-estimated the deflection; negative error value meant that the theory under-estimated the deflection. The comparison was done for before-strengthening load testing only.

Table 9.4 Theoretical and experimental deflections for bridge X-596

Truck stop	2	2	4	4	5	5
Girder	Exterior	Interior	Exterior	Interior	Exterior	Interior
Measured deflection	0.125	0.147	0.151	0.125	0.108	0.147
Mass-section analysis	0.201		0.176		0.176	
Barrier wall 0%	48%		28%		38%	
Mass-section analysis	0.164		0.143		0.148	
Barrier wall 50%	21%		4%		16%	
Mass-section analysis	0.138		0.121		0.129	
Barrier wall 100%	1%		−12%		1%	
FEM no barrier wall	0.195	0.229	0.279	0.195	0.158	0.225
	56%	56%	85%	56%	46%	53%
FEM with barrier wall	0.095	0.102	0.109	0.084	0.08	0.092
	−24%	−31%	−28%	−33%	−26%	−37%

Table 9.5 Theoretical and experimental deflections for bridge T-530

Truck stop	2	2	4	4	5	5
Girder	Exterior	Interior	Exterior	Interior	Exterior	Interior
Measured deflection	0.089	0.109	0.113	0.114	0.075	0.112
Mass-section analysis	0.145		0.124		0.124	
Barrier wall 0%	46%		9%		33%	
Mass-section analysis	0.117		0.099		0.099	
Barrier wall 50%	18%		−13%		6%	
Mass-section analysis	0.097		0.083		0.087	
Barrier wall 100%	−2%		−27%		−7%	
FEM no barrier wall	0.145	0.165	0.218	0.19	0.11	0.161
	63%	51%	93%	67%	47%	44%
FEM with barrier wall	0.093	0.135	0.14	0.145	0.075	0.135
	4%	24%	24%	27%	0%	21%

9.4.2 Results of Analysis

Tables 9.4–9.7 list the results of theoretical and (pre-strengthened) experimental deflection comparisons for each bridge except Bridge Y-298. The percentage error relative to the measured value was listed for every model. For each given truck stop, consistencies were noticed both between the models and between the bridges. Differences in percent error from one model to another were similar from one bridge to the next. When the barrier wall was considered, the stiffness increased, and as a result the estimated deflection decreased. This was true for the Mass-Section Models and the FEM Models. To quantify a health rating of the bridge, the percent errors were averaged into one number for each bridge.

Initially, it was noted that for Bridge P-962, the average percent error of the models was significantly different than that of the other three bridges (X-596, T-530, and X-495). In fact, the theoretical deflection was typically much less than the measured

Table 9.6 Theoretical and experimental deflections for bridge X-495

Truck stop	2	2	4	4	5	5
Girder	Exterior	Interior	Exterior	Interior	Exterior	Interior
Measured deflection	0.088	0.105	0.130	0.097	0.091	0.127
Mass-section analysis	0.2		0.176		0.171	
Barrier wall 0%	107%		55%		57%	
Mass-section analysis	0.164		0.143		0.143	
Barrier wall 50%	70%		26%		31%	
Mass-section analysis	0.138		0.121		0.121	
Barrier wall 100%	43%		7%		11%	
FEM no barrier wall	0.201	0.242	0.295	0.212	0.173	0.245
	128%	130%	127%	119%	90%	93%
FEM with barrier wall	0.09	0.096	0.104	0.082	0.078	0.089
	2%	–9%	–20%	–15%	–14%	–30%

Table 9.7 Theoretical and experimental deflections for bridge P-962

Truck stop	2	2	4	4	5	5
Girder	Exterior	Interior	Exterior	Interior	Exterior	Interior
Measured deflection	0.125	0.187	0.154	0.156	0.109	0.197
Mass-section analysis	0.354		0.288		0.288	
Barrier wall 0%	127%		86%		88%	
Mass-section analysis	0.140		0.114		0.114	
Barrier wall 50%	–10%		–26%		–25%	
Mass-section analysis	0.087		0.071		0.071	
Barrier wall 100%	–44%		–54%		–54%	
FEM no barrier wall	0.229	0.277	0.33	0.235	0.197	0.278
	83%	48%	114%	51%	81%	41%
FEM with barrier wall	0.134	0.206	0.17	0.18	0.095	0.203
	7%	10%	10%	15%	–13%	3%

deflection, especially with both FEM models. This indicates that a higher level of softening (i.e., cracking) had occurred with this structure. Although there were no obvious visual deficiencies in the structure, the FEM models were run again for this bridge with cracked sections instead of uncracked sections. The results are shown in Table 9.7; the percent error values for this bridge were then more consistent with the other three bridges.

9.4.3 *Inspection Ratings Comparison*

The percentage error from theory was used to quantify the health of the structure. The models were created for ideal structures, meaning that any deficiencies would not be accounted for. The limitation of these models to predict actual bridge deflections under load should be noted. Deficiencies in the structure would most likely

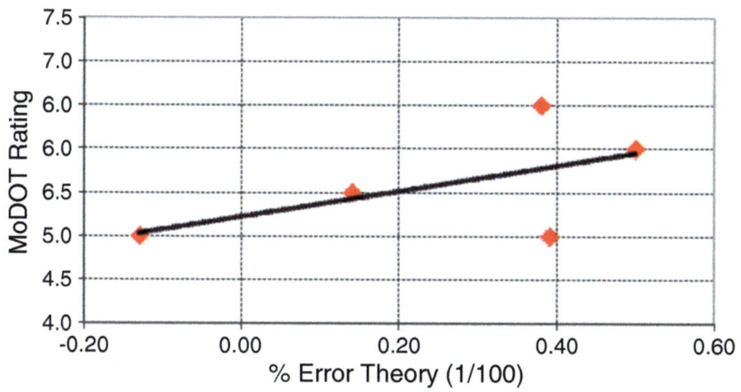

Fig. 9.15 MoDOT inspection rating vs. percent error from theory

Table 9.8 MoDOT vs. load test ratings

Bridge id	MoDOT rating	Load test rating
X-596	5.5	5.5
T-530	5.0	5.6
X-495	6.0	5.9
P-962	6.5	5.6
Y-298	5.0	5.0

reduce the structure's stiffness, increasing the deflection. Using this knowledge, we attempted to correlate the average percent error from theory to the MoDOT (2004) visual inspection ratings. The deck and superstructure ratings were averaged and plotted against the average theoretical error (see Fig. 9.15). A linear trend line was then plotted; the equation of this trend line (Eq. 9.2) was used to compute a load test rating (see Table 9.8).

$$(\text{Rating}) = 5.33 + 1.19 * (\text{Theoretical Percentage Error}) \qquad (9.2)$$

While the results suggest that there may be a relationship between load testing performance (theoretical percent error) and visual inspection ratings, this very limited data set does not ensure a trend. Many more similar structures should be load tested and analyzed to prove or disprove this rather simplistic approach. Validating this relationship would give bridge engineers a new tool to quantify their visual inspection ratings, as well as show the real condition and the load-carrying capacity of the structure.

Load testing new bridges, prior to opening them for service, would better establish a benchmark for future load tests, supplementing the analytical predictions that often yield high variability. The benchmark would be compared to future load test data to show any real degradation in the structure over time, comparing that to lowered visual inspection ratings.

9.5 Strengthening Effects

Bridge P-962 (see Fig. 9.14) and T-530 showed an expected small increase in *apparent stiffness* (decrease in deflections) from the initial load test to subsequent load tests post-strengthening. The term *apparent stiffness* is used by the authors because the actual contribution of FRP to the section properties is negligible. Much of the increase in *apparent stiffness* for these strengthened structures is due to the FRP's ability to reduce crack opening under load and thereby appear to enhance the stiffness of the structure. This increase in *apparent stiffness* does indicate that the FRP strengthening is highly engaged. For Bridge P-962, the *apparent stiffness* increase was the same for all three spans. The results of load test one (before strengthening) and after-strengthening for Bridge P-962 are shown in Fig. 9.14. The graph was normalized for the variables previously discussed. For both the longitudinal and transverse deflection, the consistent increase in *apparent stiffness* was nearly consistent for the entire span. Other bridges posted similar increases in apparent stiffness initially with less softening over time.

9.6 Changes in Deflections over Time

For Bridges P-962 and T-530, there was a clear increase in the normalized deflection over time, which approaches the original deflection of these bridges prior to strengthening. In order to better graphically present this change, the average of all deflected points was calculated for each test and a percent difference was then computed based upon the original Test 1 data. The results are presented in Figs. 9.16

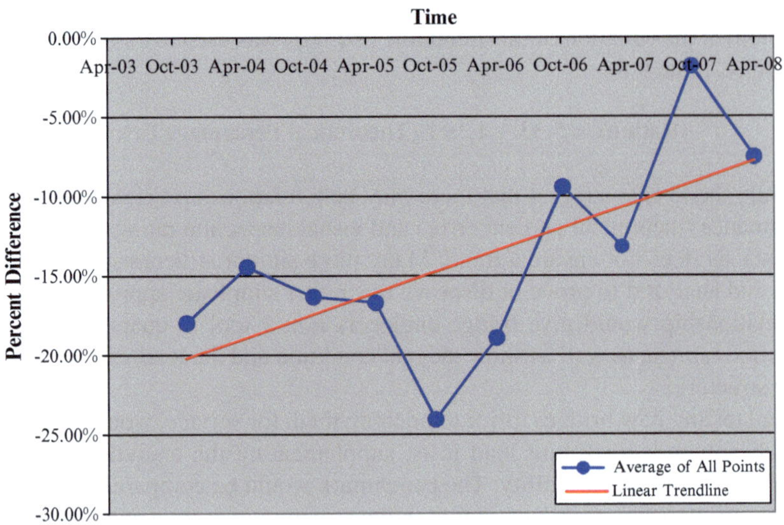

Fig. 9.16 Average deflection percent difference bridge P-962

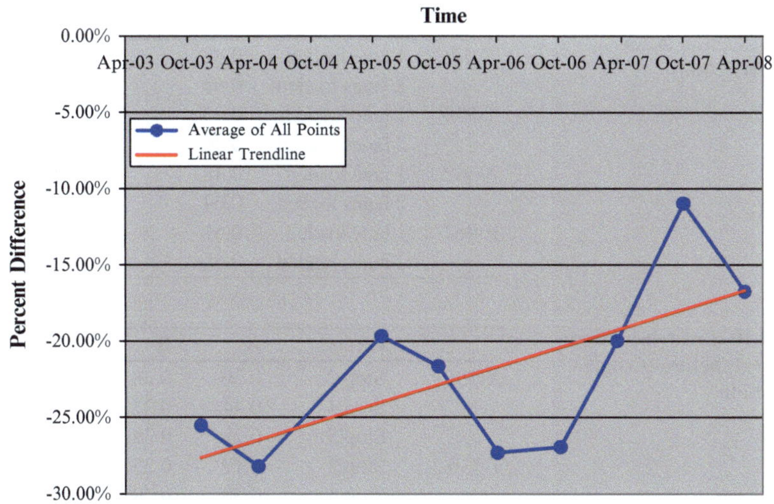

Fig. 9.17 Average deflection percent difference bridge T-530

and 9.17 for Bridge P-962 and Bridge T-530, respectively. A linearly trend line was plotted and displays a line that approaches the original deflection of the bridge. Three possibilities are considered for apparent increase in deflection: (1) bond softening at the interface of the FRP composites and the concrete substrate over the observed time, (2) softening of the FRP material utilized during strengthening period, or (3) deterioration of other bridge components (i.e., steel corrosion, concrete spalling, etc.). The varying thermal effects also may have impacted the variation and this apparent loss in stiffness due to the complexity of thermal effects with sun/shade spots during load testing, though bond softening of the laminates is suspected a primary factor based upon prior laboratory tests (Myers et al. 2001). Bridges X-495 and X-596 utilized no strengthening on the exterior girders of the tested spans. This may account for the lack of variance between the pre- and post-strengthened deflections.

9.7 Load Distribution

Live load distribution factors were attained from AASHTO LRFD Sect. 4.6.2.2 (AASHTO 2002). Table 9.9 lists the calculated distribution factors for both interior and exterior girders. The values represent the maximum response of any individual girder for a particular loading case. The case of two lanes loaded would be compared to Stop 2 of the load testing (also Stops 6 and 7 for Bridge P-962), while one lane loaded would be compared to Stops 4 and 5 of the load testing.

For every load test at each bridge, except Bridge Y-298, the individual girder response was taken from mid-span. The deflections were normalized for the variables

Table 9.9 AASHTO LRFD load distribution factors (AASHTO 2002)

		Exterior girder	Interior girder
X-596	1 lane loaded	0.48	0.60
	2 lanes loaded	0.64	0.81
T-530	1 lane loaded	0.35	0.49
	2 lanes loaded	0.46	0.64
X-495	1 lane loaded	0.48	0.60
	2 lanes loaded	0.64	0.81
P-962	1 lane loaded	0.51	0.64
	2 lanes loaded	0.68	0.86

Table 9.10 Load distribution coefficients before strengthening results

	Girder:	A	B	C	D
X-596	Stop 2	0.30	0.38	0.32	
	Stop 4	0.44	0.37	0.19	
	Stop 5	0.29	0.38	0.33	
T-530	Stop 2	0.19	0.25	0.29	0.26
	Stop 4	0.30	0.29	0.27	0.13
	Stop 5	0.18	0.28	0.32	0.22
X-495	Stop 2	0.31	0.34	0.34	
	Stop 4	0.47	0.35	0.19	
	Stop 5	0.29	0.38	0.34	
P-962	Stop 2	0.28	0.40	0.33	
	Stop 4	0.40	0.40	0.20	
	Stop 5	0.24	0.44	0.32	
	Stop 6	0.31	0.41	0.28	
	Stop 7	0.31	0.41	0.28	

Table 9.11 Percent change in load distribution coefficients from load test 1 to 2

Girder	A	B	C	D
Bridge X-596	−2	4	−1	
Bridge T-530	−1	−3	2	2
Bridge X-495	1	−3	5	
Bridge P-962	3	1	−5	

previously discussed. Again, only Stops 2, 4, and 5 were considered in the analysis. To examine the load distribution to the bridge girders, the individual girder deflection was divided by the sum of the deflections for all three girders; the result was a fraction of the load that the individual girder carries (Yang and Myers 2003). The calculated responses due to the load testing are called the load distribution coefficients. The responses used for calculation may be strain or deflection (Neely 2000). The before-strengthening responses for each bridge and each girder are listed in Table 9.10.

The load distribution coefficients ranged from 50% to 95% of the values computed from AASHTO; this implies a level of conservatism in the AASHTO LRFD distribution factors. The truck stops were set to maximize the responses in different locations instead of locating the trucks in the AASHTO design lanes. The conservative nature of the AASHTO design specification can be seen in comparing the data in Tables 9.9 and 9.10. Changes in the computed after-strengthening load distribution coefficients can be seen in Table 9.11.

Deck strengthening and flexural strengthening of different girders would produce changes in the load distribution coefficients. However, FRP strengthening only slightly altered the overall stiffness of each bridge. The small magnitude in changes from before to after strengthening, can be observed from the changes in values from Table 9.11. From the above data, it can be concluded that FRP strengthening has very little or no effect on a bridge's load distribution coefficients.

9.8 Load Testing Sensitivity

Accurately placing the H20 trucks in the load testing scheme was critical to the test's outcome. A high level of effort was undertaken to be precise in load truck measurement and axle placement throughout this study; however, a sensitivity study was undertaken to examine the influence of axle placement and load accuracy to gain a better understanding of how this might affect the load testing results. The truck's wheels were assumed to be placed within 12 in of the desired location. To find the effect of this variation, the data was modeled for Truck Stop 2 on Bridge X-596 using Finite Element Modeling (FEM), modeled without barrier walls and both trucks weighing 55 kips each. The first model was run with the loads placed at the desired locations. The second model was run with one of the two trucks simulated as being misplaced 12 in transversally. The third model was run with one of the two trucks simulated as being misplaced 12 in longitudinally. Finally, the forth model was run with one truck weighing 5 kips less than in the first model; this would simulate an error in reporting the truck weight. Truck weight was considered most at risk for error as it was not overseen by load testing personnel. Other possibilities for error included the accuracy of the surveying equipment (±0.005 in), unknown thermal and rebounding effects, different truck axle configuration dimensions, and shifted loads in the truck bed during transport. The percent changes in deflection were reported in Table 9.12. The comparison points were on each girder at mid-span. The case study for Bridge X-596 shows the sensitivity involved in load testing for this project. The errors discussed may account for the variations seen in the results discussed herein.

Table 9.12 Load testing sensitivity for an individual girder

Error	Deflection change (%)
Truck placed ± 12″ transversally	10
Truck placed ± 12″ longitudinally	5
Truck reported weight ± 5 kips actual weight	7
Surveying equipment	4

9.9 Visual Bridge Inspection

An effort was made during each of the last four series of bridge tests to visually inspect the bridges for any degradation. The structural defects observed were primarily related to the structural members of the pre-strengthened bridge which exhibited minor problems observed with the FRP materials.

9.9.1 Bridge T-530

Bridge T-530 showed signs of severe corrosion of the bents due to the propagation of water from the deck surface. These bents were corroded before the bridge was strengthened and need to be monitored in the future to ensure structural integrity. The presence of a flexural crack was also noted on the mid-span transverse beam of the tested span. The size of the crack was noted, and did not noticeably increase in size when testing under truck load. Figure 9.18 shows the location and a depiction of the corrosion of the bents and the flexural crack of the transverse girder.

9.9.2 Bridge X-596

Bridge X-596 also exhibited bent corrosion. The exterior face of one bent had spalled off, exposing the reinforcement. This occurred between tests 8 and 9 and was not

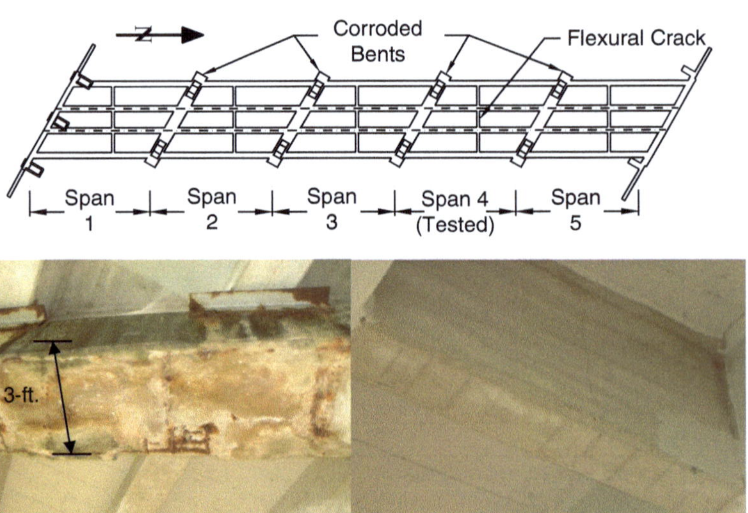

Fig. 9.18 Bridge T-530 corrosion of bents and cracked transverse beam

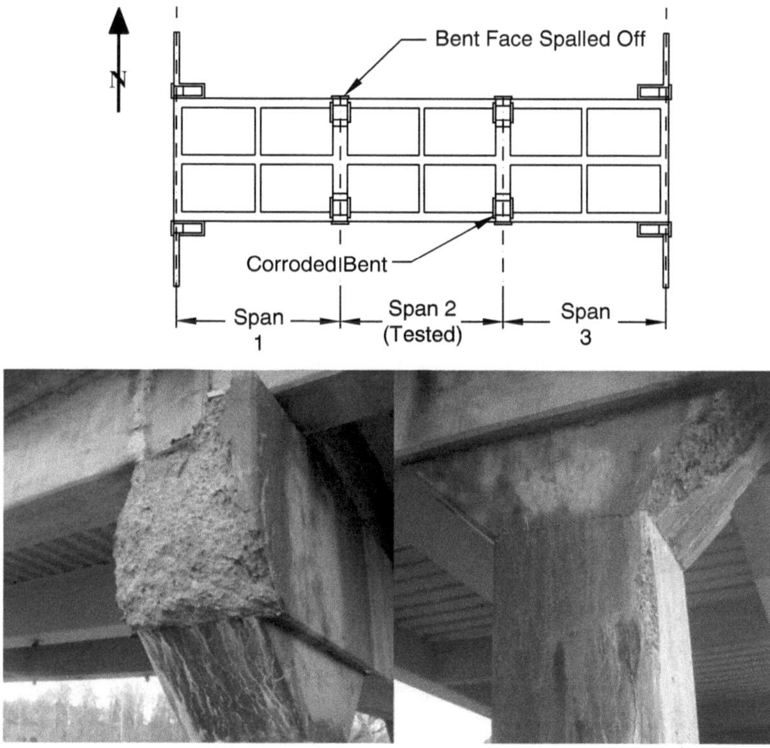

Fig. 9.19 Bridge X-596 corrosion of bents

present prior to construction. In addition, another bent exhibited corrosion at the intersection of the transverse beam with the column. The locations and a picture of each decayed bent are shown in Fig. 9.19. Additional corrosion was also observed on the barrier walls in the form of exposed reinforcement and deteriorated concrete.

9.9.3 Bridge X-495

The presence of an end-span crack in the disturbed region was noted during the eighth series of tests. This crack did not propagate further when observed during later tests. Additionally, the protective coating on the FRP applied to one bent via manual lay-up is peeling from the bent soffit. This may be due to poor installation or water drainage from the deck. This phenomenon was not observed on the other bent, which was in good condition. The disturbed region crack and protective layer peeling is shown in Fig. 9.20, along with the locations of the occurrences. Lastly, there is an appreciable amount of scour around the foundations of one of the bents. This is in the deeper part of the creek and was observed during the 8th test when the water visibility was fairly clear.

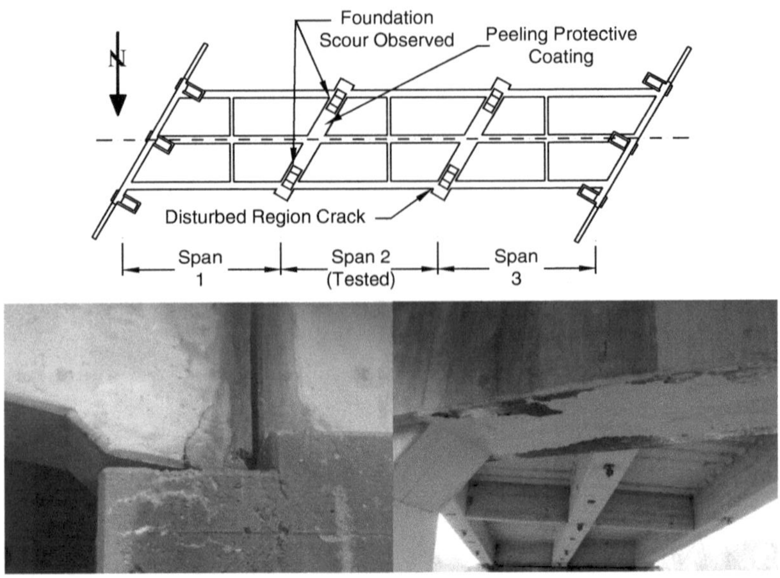

Fig. 9.20 Bridge X-495 disturbed region crack and peeling of protective UV coating

9.9.4 Bridge P-962

The transverse girder at one of the bents showed signs of corrosion in the form of exposed and rusted rebar from spalled concrete. Also, the Steel-Reinforced Polymer (SRP) utilized to strengthen the third span showed signs of rust in many places. This was especially prevalent in locations where water was able to drain from the deck to the girders or bents. The transverse girder corrosion and rusting of SRP can be observed in Fig. 9.21, along with the locations of the occurrences. The rusting of the SRP was the worst strengthening deterioration observed on any of the bridges. If SRP is allowed to rust even slightly, then the entire strengthening system could be compromised due to the overall small area of the wires. Future inspections will be very important in order to ensure the SRP is performing adequately.

9.9.5 Bridge Y-298

The condition of bridge Y-298 was relatively poor prior to the application of the FRP strengthening. For this reason, mechanically fastened FRP was utilized in deteriorated areas for quick strengthening without the tedious surface preparation required for bonded applications. The condition of Y-298 has not substantially changed since strengthening. A permanent deflection at the northern side of the eastern span still exists. Additionally, there is some scour at the eastern abutment.

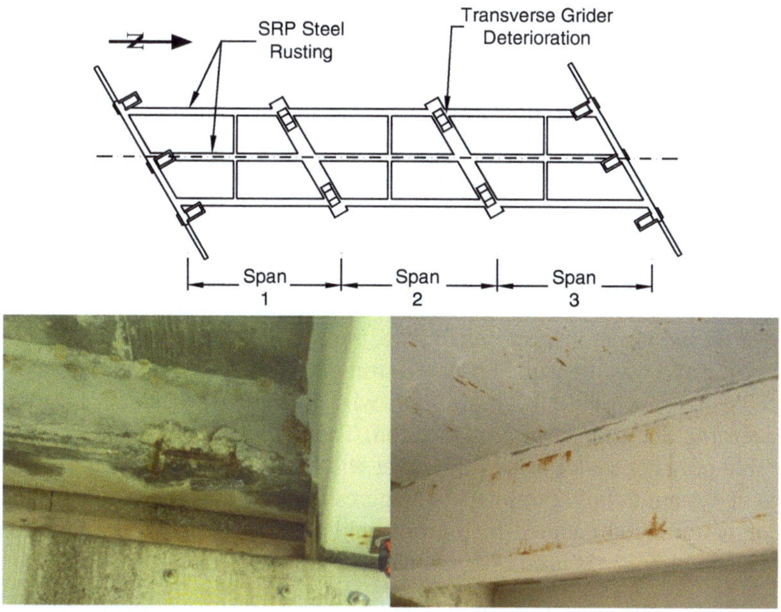

Fig. 9.21 Bridge P-962 corrosion of transverse girder and SRP reinforcement

9.10 Conclusions

The objective of this research was to evaluate the long-term service performance of the concrete bridges strengthened with FRP composites through load testing, visual inspection, and observation of any deterioration of those bridges. The following conclusions can be drawn from this research:

- Thermal effects were significant in load testing for this project due to the required time for load testing and the associated changes in the thermal behavior of the bridge throughout the load test. Shadow regions cast upon the deck during sunny days in particular made serviceability adjustments highly complex due to the varying surface and internal deck temperatures throughout the bridge.
- The apparent increase in stiffness achieved by adding FRP strengthening is primarily attributed to the restraint of concrete cracks from opening.
- There was a decrease in deflection and a subsequent apparent increase in stiffness for bridges T-530 and P-962 due to strengthening.
- For Bridges X-495 and X-596, it is difficult to quantify the apparent increase in stiffness due to strengthening with load testing. This is due to a lack of flexural strengthening of the exterior girders.
- The apparent increase in stiffness for Bridges T-530 and P-962 due to strengthening is decreasing according to an average percent difference trend line.
- The distribution fractions, or coefficients, all appeared to be within the AASHTO LRFD Load Distribution Factors. The FRP retrofit systems did not significantly

affect the load distribution coefficients when compared to the pre-strengthened values.
- Overall, there have been no changes in the deflection of the tested bridges to suggest that the FRP strengthening systems are functioning as expected.
- The visual inspections conducted during testing have yielded some important information that will be helpful in tracking future degradations of bridge components or FRP strengthening.

The following commentary is presented based on the retrofit systems implemented and general observations to date:

- Based on the visual inspections to date, the SRP system exhibited signs of steel corrosion and was the worst observed strengthening deterioration for any system used. Corrosion of the high strength wire in the SRP system is particularly concerning due to its thin cross-section. The authors believe that a much more durable polymer system is required that will better protect the steel wire for this type of system to hold promise in an aggressive external environment. This technique is not recommended for further field demonstration until a more durable protective polymer is used in combination with the steel wire.

Acknowledgements The authors would like to acknowledge that support for this project was provided by the Missouri Department of Transportation (MoDOT) and the National University Transportation Center (NUTC) at Missouri S&T. The authors would also like to thank technician and staff support from the Center for Infrastructure Engineering Studies (CIES) and the Department of Civil, Architectural, and Environmental Engineering at Missouri S&T for their assistance during the case studies.

References

AASHTO (2002) Standard specification for highway bridges, 17th edn. American Association of State Highway and Transportation Officials, Washington, DC

AASHTO (2003) Manual for condition evaluation and load resistance factor rating of highway bridges, 2nd edn. American Association of State Highway and Transportation Officials, Washington, DC

ACI 437R-03 (2002) Strength evaluation of existing concrete buildings. American Concrete Institute, Farmington Hills

ACI 440.2R-02 (2002) Guide for the design and construction of externally bonded FRP systems for strengthening concrete structures. American Concrete Institute, Farmington Hills

Benmokrane B, Masmoudi R, Chekired M, Rahman H, Debbache Z, Tadros G (1999) Design, construction, and monitoring of fiber reinforced polymer reinforced concrete bridge deck. ACI Spec Publ 188:87–102

Bridge Inspection Rating Manual (2004) Missouri department of transportation. www.modot.org. Accessed 20 May 2004

Koenigsfeld D, Myers JJ (2003) Secondary reinforcement for fiber reinforced polymers reinforced concrete panels, Center for infrastructure engineering studies, University of missouri, Rolla, No 03–45

Leica Geosystems (2004) Leica TCA 2003 total station. www.leica-geosystems.com. Accessed 20 May 2004

Merkle WJ, Myers JJ (2004a) Load distribution response of bridges retrofitted with fiber reinforced systems. Center for Infrastructure Engineering Studies, University of Missouri at Rolla, Rolla

Merkle WJ, Myers JJ (2004b) Use of the total station for load testing of retrofitted bridges with limited access. Proc SPIE Smart Struct Mater Sensors Smart Technol Civil Mech Aerospace Syst 5391:687–694

Myers JJ, Murthy S, Micelli F (2001) Effect of combined environmental cycles on the bond of FRP sheets to concrete. In: Proceedings of the international conference on composites in construction (CCC-2001), Porto, 10–12 Oct 2001, pp 339–344

Myers JJ, Holdener D, Merkle W, Hernandez E (July 2008) Preservation of missouri transportation infrastructures: validation of FRP composite technology through field testing – in-situ load testing of bridges P-962, T-530, X-495, X-596 and Y-298, Center for infrastructure engineering studies report, Missouri S&T, pp 133

Neely WD (2000) Evaluation of in-service performance of the Tom's Creek Bridge. MS thesis, Virginia Polytechnic Institute and State University, Blacksburg

Roberts GW, Meng X, Dodson AH (2004) Integrating a global positioning system and accelerometers to monitor the deflection of bridges. ASCE J Surv Eng 130(2):65–72

Stone DK, Tumialan JG, Parretti R, Nanni A (2002) Near-surface mounted FRP reinforcement: application of an emerging technology. Concrete UK 36(5):42–44

USDOT (2003) National Bridge Inventory, U.S. department of transportation, Federal highway administration, Washington, DC

Vurpillot S, Krueger G, Benouaich D, Clement D, Inaudi D (1998) Vertical deflection of a prestressed concrete bridge obtained using deformation sensors and inclinometer measurements. ACI Struct J 95(5):518–526

Watkins SE (2004) Preservation of missouri transportation infrastructure: validation of FRP composite technology through field testing – volume I. Center for Infrastructure Engineering Studies, Rolla

Yang Y, Myers JJ (2003) Live load test results of Missouri's first high performance concrete superstructure bridge. In: Proceedings for the transportation research board 82nd annual meeting, Washington, DC

Chapter 10
The Utilization of Recycled Thermoplastic Composites for Civil and Military Load Bearing Applications

Thomas J. Nosker, Jennifer K. Lynch, and Richard G. Lampo

Abstract Long-term performance and extended service life are issues of vital importance to the Department of Defense (DoD). The DoD seeks alternative construction materials to replace more traditional materials, such as wood and steel, for heavily loaded infrastructure to combat this expensive corrosion and bio-degradation problem. Recently, two military bridge installations were completed, composed entirely of a reinforced thermoplastic composite lumber (RTCL) material that is capable of supporting the load of an M1 Abrams tank at approximately 64,410 kg (71 tons). The RTCL material selected for these applications is polypropylene (PP) coated fiberglass blended with high-density polyethylene (HDPE). Advantages of using RTCL include the following qualities: corrosion, insect, and rot resistance; no toxic chemical treatments required to increase service life; environmentally friendly; diversion of waste plastics from landfills; reduction of deforestation, green house gases, and global warming. RTCL has many advantages but does behave differently than traditional materials and certain properties must be addressed during the design stage. Both bridges are continually monitored, have performed well over the first year and a half, and are more cost-effective than any other construction material. Details of the material, design considerations, and construction are reviewed.

Keywords Reinforced thermoplastic composite lumber • Recycled plastics • Polymer blends • High density polyethylene • Polypropylene • Fiberglass • Corrosion • Structural • Infrastructure • Bridge • Creep • SEET • Military demonstration project

---

T.J. Nosker (✉) • J.K. Lynch
Materials Science and Engineering Department, Rutgers University, Piscataway, NJ 08854-8065, USA
e-mail: tjnosker@rci.rutgers.edu; jklynch@rci.rutgers.edu

R.G. Lampo
U.S. Army Corps of Engineers, Engineer Research and Development Center, Construction Engineering Research Laboratory (ERDC-CERL), Champaign, IL 61826-9005, USA
e-mail: richard.g.lampo@usace.army.mil

R. Jain and L. Lee (eds.), *Fiber Reinforced Polymer (FRP) Composites for Infrastructure Applications*, Strategies for Sustainability, DOI 10.1007/978-94-007-2357-3_10,

10.1 Introduction

High-density polyethylene (HDPE) based recycled plastic lumber (RPL) emerged in the United States marketplace in the early 1990s. RPL is an attractive substitute for natural wood because it is inherently resistant to moisture, rot, and insect-borne damage; RPL does not require poisonous chemical treatments; manufacturing RPL diverts waste plastics from landfills; use of RPL reduces deforestation and has an advantageous environmental impact. These original RPL products have similar strength values to equivalent-sized wood lumber. However, the elastic modulus (stiffness) of the original RPL products is at least an order of magnitude less than even the most common wood species used in construction, and RPL is affected by significant creep at stresses of 100 psi or less. Eventually, manufacturers produced reinforced thermoplastic composite lumber (RTCL) with higher elastic modulus and creep resistance by incorporating reinforcing agents within the RPL matrix, such as fibers, or by making immiscible polymer blend composites (IMPB). Further advancements utilizing specialized processing methods resulted in nano-structured morphologies (Fig. 10.1) of the IMPB composites with enhanced mechanical properties (Fig. 10.2), including increased toughness without sacrificing modulus or strength (in both tension and compression). In Fig. 10.1 shows scanning electron microscopy (SEM) micrographs of an IMPB composite of polystyrene (PS) and HDPE processed by (a) standard single screw extrusion resulting in a coarse microstructure and (b) specialized high compounding single screw extrusion resulting in a fine, nano-sized morphology. In Fig. 10.2 shows the corresponding stress-strain curves in tension for the PS/HDPE IMPB processed by methods (a) and (b). The significance of Figs. 10.1 and 10.2 is that processing alone can provide a fine microstructure, which yields materials with higher toughness.

The materials science advancements enable RTCL to be used in a broad array of applications, including those with strict and heavy load requirements. The first vehicular bridge, composed of a PS/HDPE RTCL with typical rectangular cross

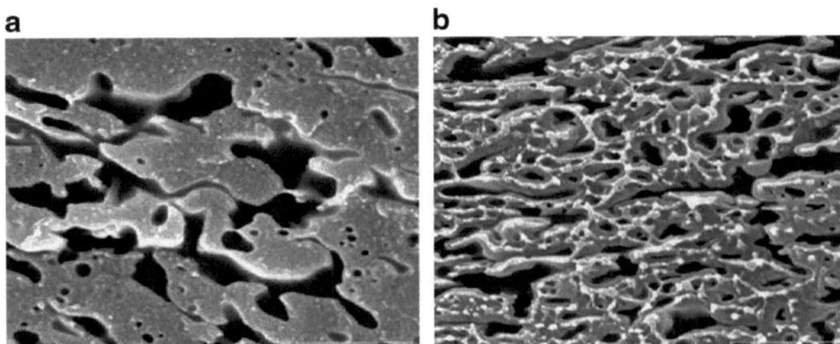

Fig. 10.1 SEM micrographs at 8.57 μm scale of a co-continuous IMPB composed of high-density polyethylene (HDPE) and polystyrene (PS) processed using (**a**) standard processing methods and (**b**) high compounding specialized processing methods (10)

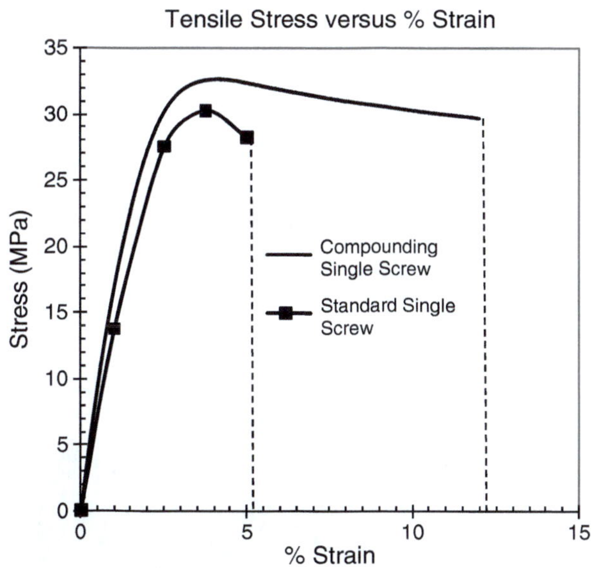

Fig. 10.2 Tensile stress-strain curves showing enhanced toughness of the nano-structured PS-HDPE IMPB prepared by high compounding specialized processing methods as compared to the same IMPB using standard processing methods (10)

Fig. 10.3 Vehicular bridge at Fort Leonard Wood, MO built in 1998 with a maximum load capacity of 12.5 tons

sections, was built at US army base Fort Leonard Wood, MO, in 1998 with an initial cost greater than comparable bridges built using chemically-treated wood (Fig. 10.3). To date, this bridge has not required any maintenance and looks new after 12 years of service. When viewed on a lifecycle cost basis, this bridge paid for its higher initial material costs in less than 8 years (2004). However, initial costs are still most often the deciding factor for material selection by facility

Fig. 10.4 Vehicular bridge in Wharton State Forest, NJ, built in 2002 with an I-beam sub-structure and a maximum load capacity of 36 tons

owners (i.e., substituting RTCL for traditional chemically treated-wood). Since 1998, researchers and engineers have sought construction designs using RTCL that are cost competitive to traditional chemically treated wood designs on a first-cost basis. Thus, arch and I-beam cross section designs were investigated as a means to reduce the material and installation costs for a given load capacity. A vehicular bridge composed of a PS/HDPE RTCL (Fig. 10.4) and I-beam sub-structure (Fig. 10.5) was built in Wharton State Forest, NJ, in 2002 with an HS-20 load rating and an initial cost comparable to an equivalent chemically-treated wood bridge. The bridge is able to support a fully loaded fire engine of 32,659 kg (36 tons). The latest demonstration of this sort is a RTCL bridge with an I-beam sub-structure with an HS-25 load rating constructed at US Army base Fort Bragg, NC. This bridge is able to support an M-1 Abrams tank (64,410 kg or 71 tons). This innovative design is cost competitive with a chemically-treated wood timber bridge designed to carry the same load. However, the RTCL I-beam sub-structure bridge is virtually maintenance-free and is not subject to degradation effects of moisture, rot, insects, weather, and corrosion (Lynch et al. 2001).

It is a significant effort to design safe, efficient RTCL structures capable of bearing heavy loads over long-term service periods with numerous load cycles, even though these are sophisticated composites. This is due to the nonlinear nature of the mechanical properties of polymers and polymer composites. For example, polymers and polymer composites are subject to creep at low stress levels, which is the permanent deformation that occurs due to a long-term constant load. The Non-linear

Fig. 10.5 I-beam sub-structure of the Wharton State Forest, NJ, bridge

Strain Energy Equivalence Theory (SEET) is a semi-empirical model that was developed and used to predict long-term creep from short-term stress-strain experiments conducted at different strain rates (Lynch et al. 2004, 2002). SEET may be used to determine the maximum allowable stress that should be generated in the bridge structure in order to avoid long-term creep. In this work, an example using stress-strain experiments in uni-axial compression and SEET is used to calculate a safe but efficient design stress for structural bridge components. This level of stress is always several times lower than the material's ultimate strength and leads to a large factor of safety to avoid fracture.

The RTCL bridge with the HS-25 load rating is the primary subject of this work. The unique features of the RTCL material, efficient structural design, and modular construction are the key elements considered by the military. Reviewed below are (1) the reasons the military sought alternative materials, (2) the bridge requirements, (3) details about the RTCL material selected, (4) design considerations, (5) construction progress, and (6) lessons learned.

10.2 Problem Statement and Project Objectives

> The insidious and pervasive effects of corrosion have now reached the point where it is a major cost for our economy and quality of life; in fact, recent studies estimate the direct cost of corrosion in the United States to be nearly $300 billion dollars per year (DoD 2009).

For the DoD alone, corrosion issues cost over $22 billion a year. According to Title 10 – Armed Forces, Section 2228 of United States Code passed in 2006, all DoD acquisitions are now required to:

> (A) ensure that the use of corrosion prevention technologies and the application of corrosion prevention treatments are fully considered during research and development in the acquisition process and
>
> (B) ensure that, to the extent determined appropriate for each acquisition program, such technologies and treatments are incorporated into that program, particularly during the engineering and design phases of the acquisition process.

Corrosion issues are of vital importance to the DoD when selecting materials for new construction and for replacement of existing structures.

> Deteriorated timber bridges are very costly to repair and more often than not, repairs can exceed the cost of replacement. Moreover, the DoD's maintenance and repair budget is not always adequate to renovate deteriorating bridges. And, if the same materials are used, the degradation cycle associated with wood material begins again. The DoD is interested in recycled plastic as a possible replacement for wood timbers at all of its military installations (Finney 2009).

Thus, the DoD requires in new, alternative materials to replace traditional chemically-treated wood and steel structures. These new types of construction materials, like RTCL, are not time-tested, as are traditional materials, and require demonstration projects and monitoring to prove their viability. The Office of Under Secretary of Defense for Acquisition, Technology and Logistics Corrosion Prevention and Control (CPC) Program and the Army Chief of Staff for Installation Management's (ACSIM's) Installation Technology Transition Program (ITTP) funded a Technology Innovation Demonstration Project at US Army Base Fort Bragg in North Carolina to construct two bridges composed of an alternative material with an HS-25 Load rating. The project was initiated by the U.S. Army Construction Engineering Research Laboratory (USACERL). ACSIM's ITTP funded the construction and initial load testing, while the CPC Program provided the funding for the design and long-term monitoring.

The primary aim of the proposed project was to construct two low-maintenance, affordable short-span bridges using recycled materials in short span applications. The project team was to construct and evaluate the structures composed of RTCL as a replacement for a conventional wood timber bridge with an HS-25 load rating (based on the American Association of State Highway and Transportation Officials (AASHTO)) and designed to support 66,224 kg (73 ton) tracked vehicles and 79,832 kg (88 ton) wheeled vehicles. Secondary aims included evaluation of the mechanical performance of the bridge, determination of the costs to complete each bridge and draw comparisons with a wooden structure of the same span and loads, and determination of other benefits of utilizing these types of construction materials and methods.

10.3 Technical Requirements

The completed bridges must comply with North Carolina State Department of Transportation (NCDOT), Division of Highway, Bridge Policy (2000 Edition) and NCDOT Division of Highway, Standards Specifications for Roads and Structures

Table 10.1 Maximum load rating

Load rating for track vehicle	73 tons	66,224 kg
Load rating for wheeled vehicle	88 tons	79,832 kg

Table 10.2 Material mechanical property requirements

Ultimate tensile stress	3,000 psi	20.68 MPa
Tensile modulus of elasticity at 1% strain (loaded at 50%/min)	350,000 psi	2.41 GPa
Ultimate compressive stress	3,500 psi	24.13 MPa
Flexural stress at 3% outer fiber strain	2,500 psi	17.24 MPa
Allowable flexural stress	600 psi	4.14 MPa
Allowable shear stress	350 psi	2.41 MPa

Standard test methods: Tensile (ASTM D638), Compressive (ASTM D695), and flexural (ASTM D6109)

All mechanical properties listed are recorded at room temperature (23°C or 73°F) and are the property minimum required

(February 10, 2006). The bridges must be composed of recycled plastics, require little maintenance over 50 years, be cost-competitive on a first cost basis with a wooden timber bridge carrying the same load, satisfy the maximum loading conditions listed in Table 10.1, and the materials specifications listed in Table 10.2.

10.4 Approach

The research team designed, built, and load tested two bridges, hereafter referred to as T-8518 and T-8519, which satisfied all requirements. The bridges were designed by McLaren Engineering Group located in West Nyack, NY, using traditional timber bridge design methodology adjusted to accommodate the unique material properties of the RTCL material. The material selected is a RTCL composed of PP coated fiberglass (FG) blended with HDPE, (FG+PP)/HDPE (Nosker and Renfree 1998, 1999), and is manufactured by Axion International, Inc. The Army Bridge Inspection Team (ERDC-GSL), along with contract support from Bridge Diagnostics, Inc., performed initial load testing to establish the Military Load Class for the bridges. Continuous, long-term monitoring of Bridge T85-18 is being conducted by Engineering Monitoring Solutions. Measurements are taken remotely from strain and deflection gauges, as well as video imagery to record the type of vehicle crossing the bridge relative to the recorded data. The 1-year report published in 2010 indicates that the bridge has performed well and has suffered no deterioration (Commander and Diaz-Alvarez 2010).

10.4.1 Material Selection

Material selection is an important consideration for any structure expected to sustain required loads. Traditional construction materials (wood, concrete, and steel) are time-tested and have decades, if not centuries, of data characterizing their attributes. Technical advances and developments in the area of materials science and engineering have resulted in new materials and composites based on thermoplastics, carbon fiber, glass fiber, and other fiber reinforced products. However, these materials are not time-tested to support their viability. And in almost every case, virgin materials were sourced to manufacture these composites.

There are multiple recycled, thermoplastic materials available for construction applications that provide corrosion resistance. However, in almost every case, these non-brittle thermoplastic materials creep at stresses too low for structural design and are, therefore, impractical.

The RTCL, (FG-PP)/HDPE, selected to construct bridges T-8518 and T-8519, is unique in several attributes. It is a combination of recycled HDPE, which is normally subject to significant creep at very low stress (e.g., 100 psi), and automobile bumper scrap, which is a composite composed of PP and FG. The materials are sourced from the consumer and industrial waste stream. The (FG-PP)/HDPE material properties are available in the previously referenced patents for specific percentages of (FG+PP) in HDPE, and the current lumber properties are available through the manufacturer, Axion International, Inc. This RTCL has a specific gravity of about 0.85 and is one-eighth the density of steel, but the resulting specific strength (strength/unit weight) is greater than many steels. Degradation due to natural UV direct sunlight does not exceed a rate of 0.003 in./year (Forster 1994). This slow rate of degradation makes this material superior to other material options.

10.4.2 Bridge Design and Considerations

Bridges T85-18 and T85-19 were designed to accommodate several tire-loaded vehicles according to HS-25 and the M-1 Abrams tank tracked vehicle, which each distribute loads over a certain area depending on the vehicle dimensions. The basic construction design includes rows of pilings (Fig. 10.6), with pins holding I-beam piling caps to each row (Fig. 10.7), a steel sill plate pre-drilled with holes to align girders (Fig. 10.8), I-beam girders spanning the length of the bridge through bolted to the pile caps and placed edge to edge across the bridge width (Figs. 10.9 and 10.10), smaller I-beams nestled in the web of and perpendicular to the larger I-beam girders, deck boards affixed to the girders with standard deck screws (Fig. 10.11), and curbing and a railing affixed to the edges of the bridge (Fig. 10.12). The width of both bridges is 16 ft and 6 in, while the spans of T85-18 and T85-19 are 42 ft and 56 ft, respectively. Dimensions for the bridge components are shown in Table 10.3.

Since the RTCL is a durable material and is not subject to moisture, rot, or corrosive degradation, the primary design considerations when using a RTCL type

Fig. 10.6 Three rows of five RTCL pilings each on T85-18 with wooden framework for construction purposes

Fig. 10.7 Pile caps in place and pinned on three row of pilings on T85-18

Fig. 10.8 Installation of one inch steel sill plate on pile caps with pre-drilled holes to serve as a template for girder installation

Fig. 10.9 Installation of girders on pile caps on T85-18

Fig. 10.10 Girders installed on T85-18 spanning the length of the bridge

Fig. 10.11 T85-18 showing deck boards spanning the width of the bridge and the railing system

Fig. 10.12 Side of deck, showing profile of bridge with railing detail

Table 10.3 Dimensions of bridge components

Component	True dimensions
Pilings	D=12 in., L=45 ft
Pile caps	18 in.×18 in.×42 ft
	Central flange=5 in.
	Top and bottom flange=3 in.
Girders	18 in.×18 in.×42 ft
	Central flange=5 in.
	Top and bottom flange=3 in.
Small I-beams	12 in.×12 in.
Deck boards	3 in.×12 in.×16 ft-6 in.
Curbing	7 in.×9 in.

of material is to account for creep. Again, creep is the residual deflection due to a constant load over a long time period after the load is removed and must be minimized by the design and choice of material. Because RTCL is not time-tested, as are traditional materials, and because long period testing is costly and inefficient, theoretical methods must be used to predict long-term creep. The Non-Linear Strain Energy Equivalence Theory (SEET) is a correlative, semi-empirical method that uses two short-term stress-strain experiments conducted at two different constant strain rates to predict long-term creep (Lynch et al. 2004, 2002). SEET has been accepted by the US automotive industry (Fan and Novak 2006) and by ASTM Sub-committee D20.20.01 Manufactured Recycled Plastic Lumber and Shapes.

SEET provides the stress for a material, below which creep is not a concern. The subsequent design of a structure may be designed so that stresses in its components never exceed the predicted stress.

Using SEET, we determined that the (FG+PP)/HDPE RTCL material is creep resistant at a stress level of 600 psi (4.14 MPa) after 25 years. In other words, the material may be subject to a stress of 600 psi over a period of 25 years, and after load removal, the material will return to its original dimensions. This includes regions where the structure observes shear, compression, tension, and flexural stress modes. Bridges T85-18 and T85-19 were designed so that stresses in the bridge components would never exceed 600 psi when loaded with 71 tons, or the M1 Abrams tank. A 600 psi stress is a small fraction of the ultimate strength of the (FG+PP)/HDPE RTCL material, unlike the method used for steel structures, in which the design stress is a much larger fraction of the ultimate strength. However, this design ensures creep management of the bridge structures since it is a demonstration project.

Once a stress value is known for a particular material, the design parameters of the structure are determinable. For example, the necessary dimensions of the components may be calculated by considering the design stress and the bending resistance of the beam, EI, in which E is the modulus of the material and I is the moment of inertia. The design team calculated that an 18×18 in cross-section I-beam would provide a moment of inertia over 6,700 in^4, which provides a sufficient EI value to obtain necessary and desired safety margins for a bridge with a HS-25 load rating. The 18 by 18 in I-beams are used as the pile caps and girders in bridges T85-18 and T85-19. The design requires 12 ft between pile caps and girders placed perpendicular to the pile caps touching edge to edge across the width of the bridge. Placing girders close together enables load sharing between girders and a high maximum capacity load rating.

Another important design consideration is thermal expansion. The thermal expansion coefficient of the (FG+PP)/HDPE material is 0.0000282 in/in/°F. For a 50 ft-long bridge with a temperature differential of 100°F, the potential change in length of the bridge is 2 in. The bridge design allows the bridge abutments to be slotted with 3 in of clearance at each end (Fig. 10.13) while the pile caps are fixed by bolts directly to the girders. This is necessary because the pile caps located at the abutments are buried in the sand and are much less free to move than non-abutment pile caps.

10.4.3 Bridge Installation

Construction methodology was typical of a timber bridge, with exceptions for the properties and quantity of materials. Standard installation equipment, construction tools, and hardware were utilized.

Pilings were driven to a 37.5 ton point refusal in order to withstand the dynamic loading of an HS-25 bridge using a vibratory hammer and a 4,000 pound diesel-powered hammer on bridge T85-18 and using a 1,000 pound bell hammer followed by the vibratory and 4,000 pound diesel-powered hammer on bridge T85-19 (Refer to Figs. 10.14–10.16). For bridge T85-18, the pilings penetrated the soil to a depth

Fig. 10.13 Bridge Abutment, showing thermal expansion slots in girders

Fig. 10.14 A crane dropping the pile into place

Fig. 10.15 Vibratory hammer driving pile

Fig. 10.16 Four-ton hammer driving piling to 37.5 ton limit

of 30 ft when the 37.5 ton refusal limit was achieved. For bridge T85-19, the soil borings indicated that refusal might not occur until almost 60 ft. Thus, the pilings were spliced with a schedule 40 steel pipe according to NCDOT regulations and driven to about 60–65 ft (Fig. 10.17). Three pilings were installed in a row at each abutment and four pilings were installed at the central location. This was a very

Fig. 10.17 Spliced pilings to achieve the depths needed for 37.5 ton refusal on bridge T85-19

important and critical consideration, ensuring that the entire structure would support very high concentrated loads across the bridge surface. Finally, the pilings were measured level and cut (Fig. 10.18).

The large I-beams serving as pile caps were affixed to the pilings with smooth stainless steel rods (1 in. diameter, 28 in length) driven into an interference fit and capped by a stainless steel washer TIG welded on to make a cap (Fig. 10.19). Stiffening boards (3 in × 12 in) were affixed to the inside of the pile caps at 18 in intervals with screws and glue (Fig. 10.20). A one-inch thick steel sill plate was placed between the pile cap and girder to serve as a template for construction with precision holes pre-drilled and to distribute stress over a larger area of the pile cap (Fig. 10.8). Each plate was carefully placed on each pile cap using laser sighting. Girders were affixed to the pile caps with four one-inch bolts per joint, and joints at each end were slotted (1 in × 4 in) to allow for thermal expansion (Fig. 10.13).

The large I-beam for the girders and pile caps were molded from two offset T-beams 42 ft in length and attached with an epoxy adhesive and then bolted together (Fig. 10.21). For bridge T85-18, this was a sufficient length for the girders to span the bridge length from abutment to abutment. For bridge T85-19, the bridge span was greater than 42 ft, and the girders were staggered so that one-third of the girder would have a splice over each non-abutment pile cap (Fig. 10.22). Between the pile caps, stress is shared by a combination of blocking between the girders at 1/3 spacing, 3 in × 12 in decking, stiffener boards below the girders also at 1/3 spans, and curbing securing the edges of the decking to the end girders.

Fig. 10.18 Trimming piling to be level

Fig. 10.19 Hammering pin through pile cap and into pile

Fig. 10.20 3 by 12 in. stiffeners installed every 18 in. in the web of the pile cap I-beam

Fig. 10.21 Girder and pile cap I-beam (18 in × 18 in × 42 ft) molded as two T-beams that are glued and bolted together

Deck boards were installed perpendicular to the girders with M5 zinc plated steel deck screws. The deck board length of 16 ft, 6 in is the bridge width. The 3 in × 12 in boards were also used to finish the abutment (Fig. 10.23). The rub railing, or curbing, was bolted to the edge of the girders perpendicular to the decking. The railing system was affixed to the bridge edge with lag bolts.

Fig. 10.22 Sub-structure of T85-19 showing the 42 foot girders staggered at one-third of the girder length to achieve a bridge span of 56 ft

Fig. 10.23 Abutment of T85-18 finished with 3 in × 12 in boards

10.4.4 Structural Health-Monitoring

Periodic inspection of the bridges is performed by local facilities personnel and contractor personnel. More detailed monitoring is being conducted under a separate task order, Degradation and Monitoring of Thermoplastic Vehicular Bridge (W9132T-06-D-0001-0055). Sensors were installed at multiple locations on each bridge to record strain, displacement, and temperature every hour. In addition, the same measurements are recorded whenever a vehicle of sufficient load passes over the bridge. This future monitoring will provide further validation of the RTCL material in high load capacity bridges, as well as lessons learned to aid in design and construction improvements (Commander and Diaz-Alvarez 2010).

10.5 Results

Construction of the RTCL bridges, T85-18 and T85-19, was successful and both performed successfully in load tests. Load testing occurred on June 11, 2009, when an M-1 Abrams tank safely crossed T85-18 and verified the safety validation of the new structure. Bridge T85-18 was dedicated on September 18, 2009, during which there was another tank crossing (Fig. 10.24). Field testing and data acquisition on the bridges post-construction confirmed that the stresses under tank loadings

Fig. 10.24 M-1 Abrams tank (71 tons) crossing the RTCL bridge T85-18 at Fort Bragg on September 18, 2009

Fig. 10.25 Loosened and dislodged deck boards on bridge T85-19 after the bridge was traversed by a 30 ton steam roller

(71 tons) were below the creep stresses of the material, and the deflection was limited to about 0.5 in. This deflection is well below the allowable deflection, according to the bridge designs. Interestingly, the pile deflections under load for the two bridges were similar, even though there were some installation differences. The girders spanned the length of the bridge on T85-18 but were spliced to accommodate a longer bridge span on T85-19, and the pilings for T85-19 were driven to about twice the depth of the pilings for T85-18. This indicates that the skin friction between the piles and the soil for T85-19 piles, driven to deeper soil depths, was significant enough to lend significant support for the bridge.

In addition to the M-1 Abrams tank, several other types of vehicles crossed the bridges, including a 30 ton steam roller. The steam roller applies 300 lbs/linear inch along the contact between the roller and the deck surface, resulting in a load applied to only one deck board at a time rather than an area. The steamroller was not listed in the potential vehicles to cross the bridges and this type of almost point loading was not accounted for in the design considerations. This was immediately apparent when the steam roller crossed T85-19. Some of the deck boards on the outer edges of the bridge bent upward, as the screws located on the few girder-deck board connections were partially dislodged (see Figs. 10.25 and 10.26). This type of damage between the deck surface and the girders decreases the potential load sharing between the girders and increases the stresses due to the applied load directly under the wheels. This is in contrast to bridge designs that attempt to optimize the deck board-girder connections to help lower the maximum stresses imparted during loading.

Fig. 10.26 A deck board dislodged 0.5 in above the deck surface on bridge T85-19 after the bridge was traversed by a 30 ton steam roller

Modifications were made on both bridges to ensure future successful crossings of the 30 ton steam roller (Fig. 10.27). On T85-19, the dislodged screws were removed and replaced with longer screws, and additional screws were installed so that the four outermost girders on each side of the bridge have nine screws at each deck board-girder connection. On T85-18, an additional three screws per deck board-girder connection were installed on the outermost four girders on each side of the bridge. At this time, a rub rail or curbing was bolted through the decking to the girder on both bridges to provide additional support between the decking and outermost girder and to provide a contact other than the rails and posts for a vehicle that is too close to the edge of the bridge. Subsequent crossings of the steam roller after the modifications on both bridges were successful.

10.6 Discussion

RTCL timbers were used for the pilings, pile caps, girders, decking, and railing system to complete bridges T85-18 and T85-19 at Fort Bragg, NC, with a maximum load capacity sufficient to handle an M-1 Abrams tank (71 tons). They are the highest load capacity bridges known to the authors that are composed from thermoplastic polymer, recycled post industrial and consumer materials. Both bridges used over 85,000 lbs of recycled plastics each. Innovative plastic I-beam components

Fig. 10.27 A successful crossing of a 30 ton steam roller over bridge T85-19 after repair of the decking

were used to support heavy loads and to provide a design that compares favorably in cost to a treated-wood bridge of the same load capacity and span.

In fact, the cost/ft^2 to build a RTCL bridge like T85-18 was approximately 32.9% lower than a traditional wood timber bridge and 54.8% lower than a steel/concrete bridge designed to carry the same load. These costs are derived from bid awards and actual contracts for bridges of a similar rating (HS20-25) and equalized as appropriate for the load rating. The steel/concrete bridge is more costly due to the curing time of concrete. When poured on site, concrete must cure for 28 days. A RTCL bridge has no such limitations and is ready for use immediately.

The projected return on investment (ROI) is favorable for a bridge constructed from the RTCL materials because it is virtually maintenance-free and has a very low life cycle cost, as compared with a wood timber bridge or steel/concrete bridge. This was exemplified in the Fort Leonard Wood installation, in that the bridge paid for itself in less than 8 years due to lack of maintenance required.

Conversely, the maintenance of wood timber bridges can be significant in terms of time and expense. The following scenario is an example of the maintenance required for a typical wood timber bridge and was prepared by a former owner of a wood timber bridge building company. Consider a two-lane bridge with a HS-20-44 load rating that is 26 ft wide and has a span of 100 ft. Scheduled maintenance is required at 6 months, 2 years, every 2 years afterward, as well as major replacements after 5 years.

After 6 months, a new wood structure requires surface treatment, including pressure washing and polyurethane coating to seal the wood and protect against UV rays (Table 10.4). After 2 years, and at 2 year intervals from then on, the surface treatment

Table 10.4 Treatment costs after 6 months for an example chemically-treated wood timber structure

Treatment costs (US $)	
Pressure wash	600
Polyurethane treatment	800
Environmental protection	300
Labor (two-man crew at $500/day over 3 days)	1,500
Total	3,200

Table 10.5 Maintenance costs required for a two-lane chemically-treated wood timber bridge with a HS-20-44 load rating that is 26 ft wide and has a span of 100 ft

Years	0.5	2	4	5	5.5
Treatment costs (US $)	3,200	3,200	3,200	0	3,200
Minor repairs and replacement	0	2,900	2,900	0	0
Deck replacement	0	0	0	20,000	0
Sub-total	3,200	6,100	6,100	20,000	3,200
Total	3,200	9,300	15,400	35,400	38,600

is required, as well as minor repairs and replacement of the exposed boards. After 5 years, and every 5 years afterward, the decking is typically replaced due to wear. After 10 years, the other components may require replacement, including the guardrail, handrail, and curbing. Refer to Table 10.5 for a detailed summary of maintenance costs associated with an example chemically-treated wood timber bridge over the first 5 years. Note that the numbers provided due not account for new materials or labor during replacement. The maintenance costs of a wood timber structure are $38,600 after only 5 years, while the maintenance costs of a RTCL structure in the same location are virtually zero. The savings in maintenance costs between a wood timber structure and a RTCL structure increases drastically over even longer time periods.

The Office of the Secretary of Defense sponsored a series of studies by the Logistics Management Institute (LMI) to determine the cost of corrosion to the DoD. During the bridge dedication ceremony on September 18, 2009, at Fort Bragg, Dr. Roger Hamerlinck, Senior Acquisition Policy Specialist, stated:

> The latest study shows that the DoD spends $22.5 billion annually on equipment and infrastructure as an impact of corrosion. For the Army, this number is approximately $5.8 billion annually." He also noted that, "This bridge is less expensive to build than its alternatives, it provides greater corrosion resistance, and it is practically maintenance free. The Army estimates that we will receive a 34–1 return on investment by using this technology. 10.7

10.7 Conclusions

The first two bridge installations of a recycled, thermoplastic composite bridge capable of withstanding a 71 ton load were a technological and cost-effective success. Based upon data obtained from field testing and affixed monitoring devices of bridges T85-18 and T85-19 at Fort Bragg, it is apparent that this technology is both viable and desirable for all structural short-span bridge applications. The thermoplastic composite products used for these bridges are made from virtually 100% recycled materials, they are cost competitive on an installed cost basis as compared to traditional materials, the life cycle costs are far lower than traditional materials since RTCL is virtually maintenance-free and not subject to corrosion or degradation, and construction time is much faster than traditional materials. This "rapid build" capability of the RTCL materials allows construction completion in 4 weeks for a bridge design like T85-18 or T85-19, as opposed to a steel/concrete bridge that requires significant curing times that require construction delays. The design/research team has identified methods to further reduce installation time on future bridges.

Key next steps to advance the implementation of the RTCL technology include the development of guide specifications that articulate the detailed material characteristics, as well as recommended design and construction techniques. In addition, new I-beam geometry is being explored that will enable longer spans and greater load bearing capability. These advancements will allow RTCL to be utilized in other applications, including marine bulkheads, fender walls, and tactical bridge applications. This (FG+PP)/HDPE RTCL material is light weight but has a high specific strength and lacks significant radar signature, which encourages the application of tactical bridges to be supplied to strategic sites in a modular or kit form and quickly deployed in the field.

This recycled thermoplastic technology is now ready for wide-scale deployment and must be considered for all structural applications in the future. Using this material will ensure that corrosion and degradation is no longer an accepted norm and that future generations will benefit from this significant and innovative advancement.

The Fort Bragg project received a strong vote of confidence from the DoD due to its success, as stated by Daniel J. Dunmire, Director of Corrosion Policy and Oversight:

> This thermoplastic bridge, able to withstand heavy loads with little to no maintenance, expected to last at least 50 years, is no longer the bridge of the future – it's the bridge for today. It also meets national environmental goals of being completely recyclable. This technology is not only good for DoD, it should be immediately transferred to state Departments of Transportation for use wherever possible.

Acknowledgements The Authors would like to acknowledge Rutgers University, US Army CERL, and Axion International, Inc., for their involvement in the pursuit of the technologies discussed in this book chapter.

The authors also recognize and thank the Programs, Offices, and Sponsors that funded and supported the design, construction, and testing of the thermoplastic composite bridge at Fort Bragg:

- Office of Under Secretary of Defense, Office of Corrosion Policy and Oversight (Director, Dan Dunmire).
- Deputy Assistant Secretary of the Army Acquisition Policy and Logistics (Army Corrosion Control Prevention Executive, Wimpy D. Pybus).

- Assistant Chief of Staff for Installation Management (David Purcell, CPC Program; Philip Columbus, ITTP Program Manager; and Michael Dean).
- Headquarters, U.S. Army Installation Management Command (Paul Volkman, CPC Program, and Ali Achmar).

The authors would also like to recognize the team that made the thermoplastic composite bridge at United States Army Base Fort Bragg, NC, a reality and success. We thank them all for their support, dedication, and professionalism.

- Kelly Dilks, ITTP Project Manager and Vincent Hock, Army Facilities CPC Program Project Manager of the Engineer Research & Development Center (ERDC), Construction Engineering Research Laboratory (CERL)
- Vincent Chiarito, Terry Stanton, and Henry Diaz-Alvarez of the ERDC, Geotechnical and Structures Laboratory (GSL)
- Gregory Bean (Director), Darryl Butler, George Whitley, Fred Plummer, Lowell Stevens, and Robert Gardner of the Fort Bragg Directorate of Public Works
- Malcolm McLaren and George Assis of McLaren Engineering Group
- James Kerstein of Axion International, Inc.
- Larry Clark and Karl Palutke of Mandaree Enterprises Corporation

References

Commander B, Diaz-Alvarez H (2010) Field testing and load rating of the world's first thermoplastic bridge. ERDC/GSL TR-10-19, Vicksburg, pp 1–105

DoD (2009). http://www.corrdefense.org/CorrDefense%20WebPage%20Content/CorrosionAPersistentBattle.aspx. CorrDefense Corrosion Document Management

Fan B, Novak GE (2006) Creep prediction for polymer: implementation and its application on a talc filled polypropylene, SPE-ANTEC technical papers

Finney D (2009) Recycled plastic bridge at Fort Bragg stands up to M-1 traffic, DPW Digest

Lynch JK, Nosker TJ, Renfree RW, Krishnaswamy P, Francini R (2001) Weathering effects on mechanical properties of recycled HDPE based plastic lumber, SPE-ANTEC technical papers

Lynch JK, Nosker TJ, Renfree RW (2002) Polystyrene/Polyethylene composite structural materials, the center of advanced materials via immiscible polymer processing

Lynch JK, Van Ness K, Nosker TJ, Renfree RW (2004) Creep prediction using the non-linear strain energy equivalence theory, SPE-ANTEC technical papers 1927–1931

Nosker TJ , Renfree RW (1998) US Patent 5,789,477, Composite building materials from recyclable waste, issued Aug. 4, 1998

Nosker TJ, Renfree RW (1999) US Patent 5,916,932, Composite building materials from recyclable waste, issued June 29, 1999

Ronald George Nathaniel Forster (1994) The ultraviolet degradation of polyethylene and poly(ethylene carbon monoxide) under simulated environmental conditions, Rutgers University

Chapter 11
Structural Response, Health Monitoring, and Performance Evaluation of CFRP Post-Tensioned, In-Service, Long-Span, Precast/Prestressed Box Girder Bridges

Manoochehr Zoghi, Patrick Plews, and Dean C. Foster

Abstract Two severely deteriorated long-span, precast, prestressed, box-girder bridges, located in Defiance, Ohio, were selected by the Federal Highway Administration's (FHWA's) Innovative Bridge Research and Construction Program for rehabilitation using advanced carbon fiber polymer (CFRP) composite materials, full-scale live-load testing, and long-term structural health monitoring. The root cause of deterioration was an improper deck drainage scheme, which had enabled the de-icing salts and water to corrode the prestressing strands and spall the concrete. Sika's CFRP stress-head system, a post-tensioned and bonded external strengthening technique, was utilized to restore the original load carrying capacity of the deteriorated beams. An integral component of the rehabilitation plan was a long-term, structural health monitoring system (SHM), installed to monitor the performance of selected beams on the bridges. The structural health monitoring system consisted of vibrating wire strain gages with integrated thermistors to monitor changes in strain and temperature on 11 beams on the bridges. The data compiled reveal that compression is induced in the bottom of the deteriorated beams, thereby improving the load-carrying capacity. The use of the external Stress Head

M. Zoghi (✉)
Lyles College of Engineering, California State University, Fresno,
2320 East San Ramon Ave., M/S EE94, Fresno, CA 93740, USA
e-mail: mzoghi@csufresno.edu

P. Plews
TranSystems, 4555 Lake Forest Drive, Suite 540, Blue Ash, OH 45242, USA
e-mail: pjplews@transystems.com

D.C. Foster
Composite & Hybrid Materials Branch, Wright Patterson Air Force Base,
Dayton, OH 45433-7750, USA
e-mail: dean.foster@wpafb.af.mil

R. Jain and L. Lee (eds.), *Fiber Reinforced Polymer (FRP) Composites for Infrastructure Applications*, Strategies for Sustainability, DOI 10.1007/978-94-007-2357-3_11,

post-tensioning system proved to be a viable and effective means to strengthen a deficient box girder bridge. In addition, a series of full-scale live load tests were conducted prior to and following the completion of bridge rehabilitations. Strain transducers were employed to monitor the structural response under the multiple passes of truck loads. An overview of the compilation of the short-term live load test results and long-term structural health monitoring data is presented in this manuscript. The behavior of both bridges in terms of strain response history is summarized.

Keywords Stress-head system • Rehabilitation • Carbon fiber reinforced polymer composite (CFRP) • Health monitoring • Post-tensioning • Performance • Vibrating wire strain gages

11.1 Introduction

It is well known that the corrosion of prestressing steel strands/tendons is a root cause of deterioration of structures in cold regions, where deicing salt is frequently used in the winter (Grace et al. 2006). Two such bridges, built in Defiance, Ohio, in 1979, experienced severe deteriorations recently, suffering from various levels of concrete spalling and reinforcing corrosion. The reference bridges, utilizing side-by-side prestressed concrete box-beam type bridge superstructure construction, had faulty deck drainage systems. The existing gap between the roadway and sidewalk beams on both bridges had enabled the deicing salts and storm water to flow onto the underside of the fascia box girders and penetrate to the level of the reinforcing bars and prestressing strands. As a result, extensive corrosion of reinforcement, coupled with widespread cracking and spalling of concrete, had threatened the structural integrity of the afflicted girders (shown in Fig. 11.1). An estimated 25% of the tendons were either entirely lost or partially damaged due to the corrosion effect.

The Hopkins Street Bridge, a six-span simply supported prestressed concrete bridge, crosses the Auglaize River, carries approximately 8,000 vehicles per day, has two lanes, with the total length of 511 ft (155.8 m). Similarly, the Clinton Street Bridge possesses six spans and is constructed of precast, prestressed concrete box beams. This bridge, however, carries 26,000 vehicles per day, has four lanes, and crosses the Maumee River. Its total length is 430 ft (131 m). The deck drainage of Hopkins Street Bridge routed the deicing salt mostly toward the fascia beams of the bridge deck, causing severe deterioration of these beams. In the case of Clinton Street Bridge, the drip strip was pointing toward the walkway beams and inflicted greater damage to the prestressing strands and concrete spalling of the walkway beams.

After thoroughly investigating several strengthening alternatives, involving conventional post-tensioning systems, passive carbon fiber reinforced-polymer composite systems, etc., it became evident that Sika's Stress Head System™ was the most viable option (refer to Fig. 11.2 for details). It is known that the reference

Fig. 11.1 Severe case of a deteriorated roadway beam, exhibiting complete corrosion of existing prestressing steel

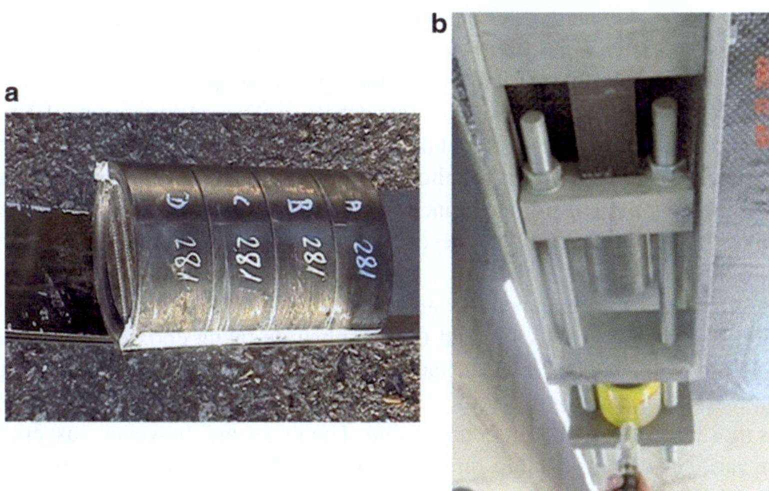

Fig. 11.2 (**a**) Stress head component of the CFRP strip. (**b**) Stressing anchor

Stress Head System combines the advantages of post-tensioning applications with the high tensile strength capacity of prestressing laminates. The stressing anchor is also constructed of stainless steel, having a basic rectangular wedge shape with a load transfer dowel near one end and at the thickest part of the wedge. The dowel has an outer diameter of 110 mm, an inner diameter of 84.6 mm, and a length of 190 mm. The stressing anchor is 785 mm long, 210 mm wide, and has a maximum height of 145 mm, as shown in Fig. 11.2. The noncorrosive nature of the fibers, the ability to retrieve strain in existing tendons as well as securing the laminates, to the repaired concrete surface via a combination of mechanical and chemical means to ensure a safe and long-lasting repair, are some of the unique features of this application. The bridges were slated for post-strengthening full scale load tests and structural health monitoring to scrutinize the stress-strain and deflection patterns. One of the primary objectives of the present study was to utilize a state-of-the-art structural health monitoring (SHM) system to investigate the long-term performance of the CFRP post-tensioned repair methods implemented on the Hopkins and Clinton Street bridges. The scope of work related to the design, construction, maintenance, data collection, and analysis for the SHM system, adopted in the current study, will be presented in this paper. In addition, a series of full-scale live load tests were conducted on both structures. A summary of structural responses and performance evaluation of bridges is discussed.

11.2 Literature Survey of SHM Systems

Kessler (2002) states: "*structural health monitoring (SHM) is an emerging technology leading to the development of systems capable of continuously monitoring structures for damage to improve safety and reduce life-cycle costs.*" By and large, SHM systems are made up of a number of subsets including data acquisition, communication, storage, processing, and retrieval of data (ISIS 2001a). Acquisition of the data begins with a system of sensors installed on the structure. There are a number of sensors available for long-term monitoring of structures, each with its own advantages and limitations. The primary concern, when monitoring the structures, involves the movement of the structure, and a secondary concern involves the environmental exposure of the structure.

Foil strain gages are inexpensive and are typically used to measure strains in a structure. They are normally bonded to the substrate using epoxies or other adhesives and measure changes in electrical resistance due to variations in the length of the thin metal foil gage. Foil strain gages are capable of detecting dynamic strains such as those due to live loading on a bridge. The gages are, however, susceptible to drift or variances from a stable reference point and are sensitive to humidity and moisture (ISIS 2001a). Furthermore, they are not very durable, failing within a few months of service.

Vibrating Wire (VW) strain gages are also commonly used in the monitoring of structures (Shepherd 1964). VW gages function by magnetically exciting a steel wire to vibrate and measuring the frequency at which it vibrates. Vibrating wire gages are larger than most other sensors used to measure strains but are significantly more durable, able to last many years. Vibrating wire sensors also include a thermistor for measurement of temperature. These gages, however, exhibit low resolution due to the time required to "pluck" the wire and take the reading. They are stable over long periods of time, reported as 0.1%FS/year (Geokon 2002).

Fiber Optic Sensors (FOS) were considered due to their small size and their ability to be easily integrated into FRP structures. The major advantages of fiber optic sensors are their immunity to interference, stability over time, high sampling rates and absolute measurement of small changes (ISIS 2001b). Fiber Bragg Grating and Fabry-Perot sensors are two types of fiber optic sensors that have been developed for monitoring strains.

11.3 Comparison of Available Systems

Traditional foil strain gages were dismissed for application on a long-term basis due to their susceptibility to drift and outdoor exposures. The fiber optic sensors and vibrating wire sensors both were found to have favorable applicability for long-term structural health monitoring. The FOS's have the advantages of being able to capture the dynamic loading effect and are malleable enough to be placed along the bond line between the Stress Head plate and the beam substrate. Fiber optic sensors were not, however, acquired in this project primarily due to lack of funding.

Long term reliability, ruggedness, ease of installation and use, integrated temperature mechanism, and watertight characteristics are some of the advantageous features of VW sensors. As a result, it was determined that vibrating wire sensors would be the most viable system for monitoring the long-term performance of the reference bridge superstructures. The low resolution of vibrating wire systems is not conducive to monitoring the effects of live load on a structure. This was the primary reason the short term full scale live load tests were performed separately, utilizing strain transducers, in collaboration with the Bridge Diagnostics Incorporated (BDI). The results of the live load tests are available in a comprehensive report (BDI 2004).

11.4 Vibrating Wire System Details

The Hopkins Street's long-term monitoring system was configured and programmed to take measurements from 48 VW strain gages and the corresponding thermistors. The data was collected using a Campbell Scientific (CSI) CR-10x datalogger. The datalogger was powered by a 12 V deep-cycle battery, and communication was accomplished via a telephone modem (CSI-COM210). All strain gages utilized vibrating wire strain gages (Geokon VK-4100) and were activated by a single vibrating wire interface (CSI-AVW1). Sensors were connected to, and read sequentially with, the use of three 16-channel multiplexers (CSI-AM16/32). The data collected were averaged hourly and stored for additional processing. The Clinton Street's long-term monitoring system was identical to the system for Hopkins Street's System, except only 32 sensors and 2 multiplexers were used.

11.5 Installation Details

The sensors were attached to the structural members using two different methods. Sensors that were mounted to CFRP reinforcement were bonded using Sika's Sikadur AnchorFix-1™, a two-component acrylic adhesive. Sensors mounted to concrete were anchored using masonry screws and a contact adhesive or the AnchorFix-1. Numerous sensors, mounted using only the AnchorFix-1, were found to have debonded. Sensors were adjusted to the mid-range setting at installation, allowing for equal ranges of tension or compression. Wiring for the sensors was run by the general contractor. Attachment of the sensors and connection of the wiring were performed by the Bridge Diagnostics, Incorporated (BDI) and researchers at the University of Dayton.

The dataloggers were mounted on the bridge abutments in weather-tight and vandal-resistant enclosures. The 12 V deep cycle battery was mounted in a separate enclosure adjacent to the datalogger. The datalogger and battery for Clinton Street Bridge were mounted near the top of the 20ft tall abutments to avoid damage from potential flooding. The higher elevation made the enclosures accessible only by a ladder. The 12 V battery supply powered the data acquisition system for approximately 2 months before recharging was required.

11.6 Data Collection Methods

Data was collected from the dataloggers using Campbell Scientific's Loggernet software. Bridge Diagnostics, Incorporated provided the additional programming within Loggernet to collect and store the data. The program was set up to operate in two different modes. The primary mode recorded the sensor readings (thermal and strain) and stored the average hourly temperature, average hourly strain, and minimum and maximum strain value per hour. A secondary mode was also setup within Loggernet that would store each sensor reading (thermal and strain) at the program cycle and also the change in strain per cycle relative to a selectable baseline reading. Data file output was in comma-separated values. Data files were transferred directly from the data logger to a laptop computer until the spring of 2004, when a telephone line was installed. Following installation of the telephone line, communications were possible using the datalogger modem and a remote computer.

11.7 Processing and Reduction of Data

The processing of the data began by importing the average hourly thermal and raw strain values into Microsoft Excel. The data for each sensor was then checked for periods of time when the sensors went out of range. The sensors typically went out of range due to the battery voltage dropping below 10.5 V. Changes in strain from the initial reading were then calculated including a thermal expansion correction. The thermal expansion or temperature correction is used to compensate for the difference in thermal expansion coefficients between the steel gage and the material it is bonded to. An example of the temperature correction is presented in the Geokon Instruction Manual (2002). The thermal coefficient of concrete was used in the temperature correction calculations, even though some sensors were bonded to FRP materials. Design of the rehabilitation system assumed that all FRP materials were fully bonded to the concrete's substrate, and therefore it was believed that the FRP materials would expand and contract with the concrete substrate to satisfy strain compatibility. The change in strain was then plotted versus time for each sensor.

The compilation of data and plots of strain versus time are included in a final report submitted to the Ohio Department of Transportation and Defiance County Engineer Office (Zoghi and Foster 2006). The highlights of that report, along with unique features of the strain time history data, are presented herein. Specifically, the seasonal data, representing the performance of CFRP plates, are of utmost significance. Figure 11.3 illustrates the plot of strain time history for a sensor that was mounted near the anchorages, with one leg on the CFRP plate and the other leg anchored to the concrete surface. This particular senor was installed right after the post-tensioning process was completed and the first 3 months of readings reveal a loss in the post-tensioning force. The change in the average hourly values is approximately 575$\mu\varepsilon$, which correlates to a 13,760 psi stress intensity in the CFRP, about 5% of the initial post-tensioning stress. This loss is attributed to a combination of creep, relaxation, anchorage set up, and shrinkage. The design of beams is typically based on the assumption of 10% loss for the anchorage set up and a 30% long-term loss for creep, shrinkage, and relaxation. It is evident that after the first 3 months, the beam exhibits a gradual increase in tension of about 150$\mu\varepsilon$ from its lowest value. This change in strain is similar to that seen on other beams and is most likely attributed to the volumetric changes caused by the efflorescence visibly observed, described below.

Figure 11.4 represents the results from a sensor that was mounted as high as possible on the side of a beam at mid-span. Although the intent was to install the sensor above the neutral axis of the beam, the limited space between adjacent beams made it difficult to reach that height. The results are indicative of a sensor going into compression when the post-tensioning force was induced, indicating that the sensor was actually located below the neutral axis. The plot also reveals that the beam undergoes a differential thermal cycle.

Figure 11.5 represents the strain time history of a sensor that was mounted on the CFRP wrap near the anchorage. This sensor generally indicates a thermal cycle; however, it does not return to the initial value after the passage of a day. Furthermore, its value 1 day later reveals that compression has been induced on the order of 15 $\mu\varepsilon$. The thermal stresses generated appear to be due to differential thermal effects on the concrete beam. An increase in compression is noted around 5 h mark when the post-tensioning force was applied.

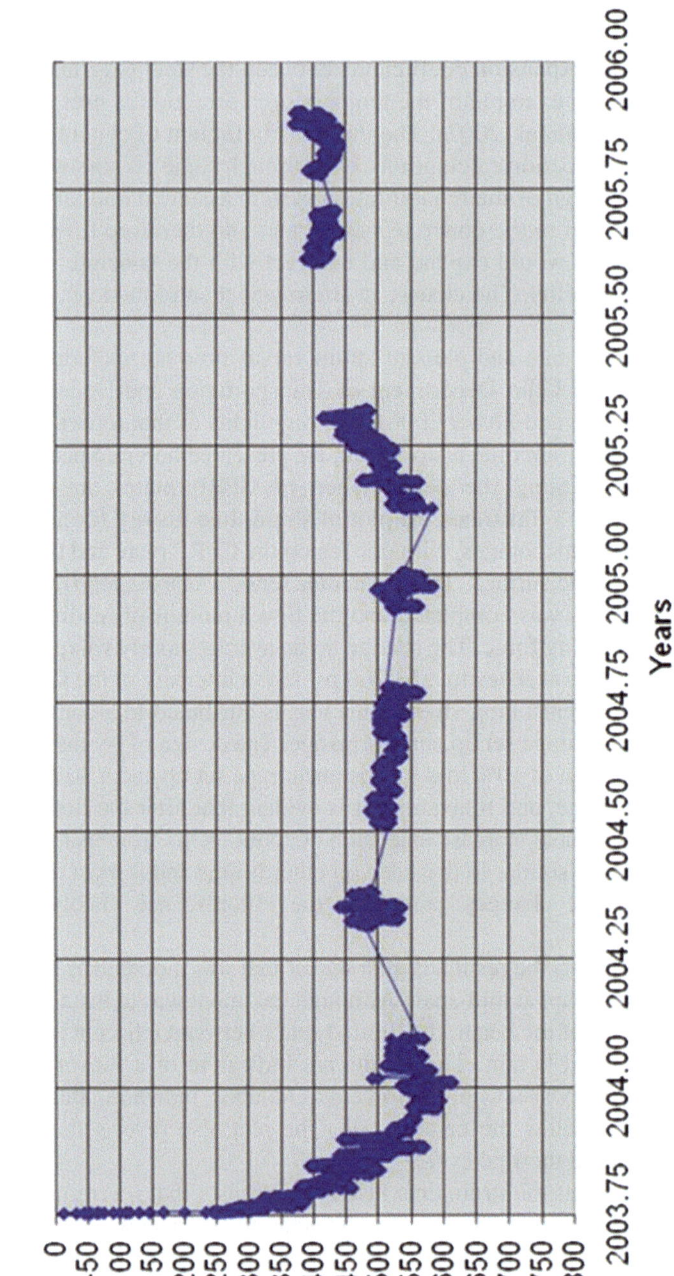

Fig. 11.3 Typical strain time history for a sensor mounted on CFRP and concrete

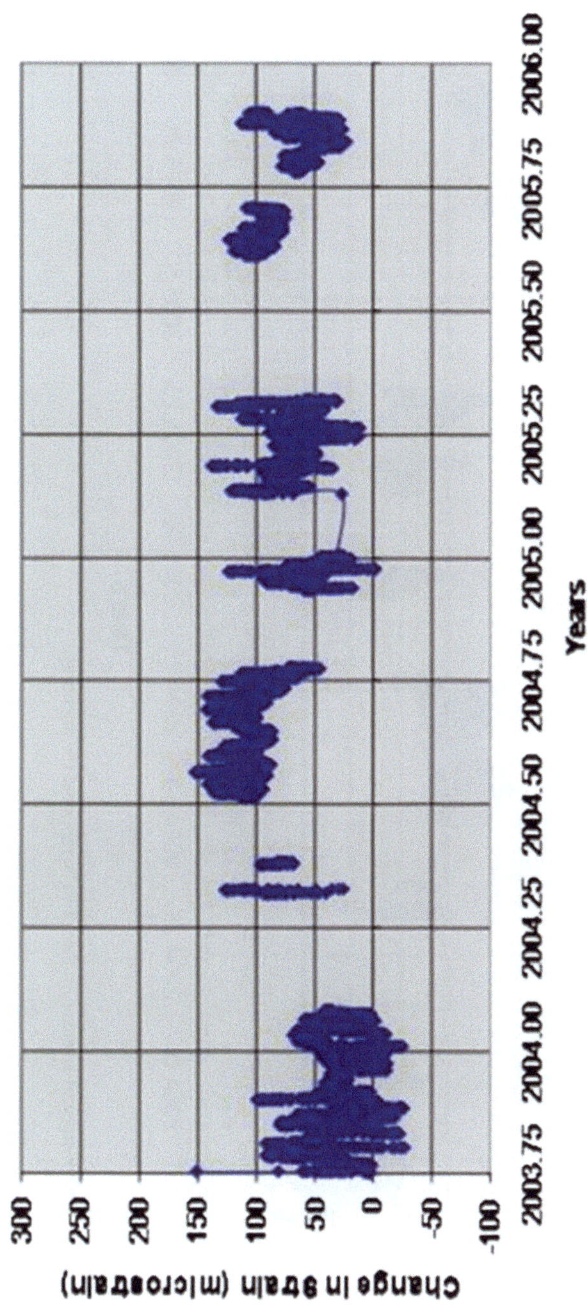

Fig. 11.4 Strain time history for a sensor mounted on the side of a beam

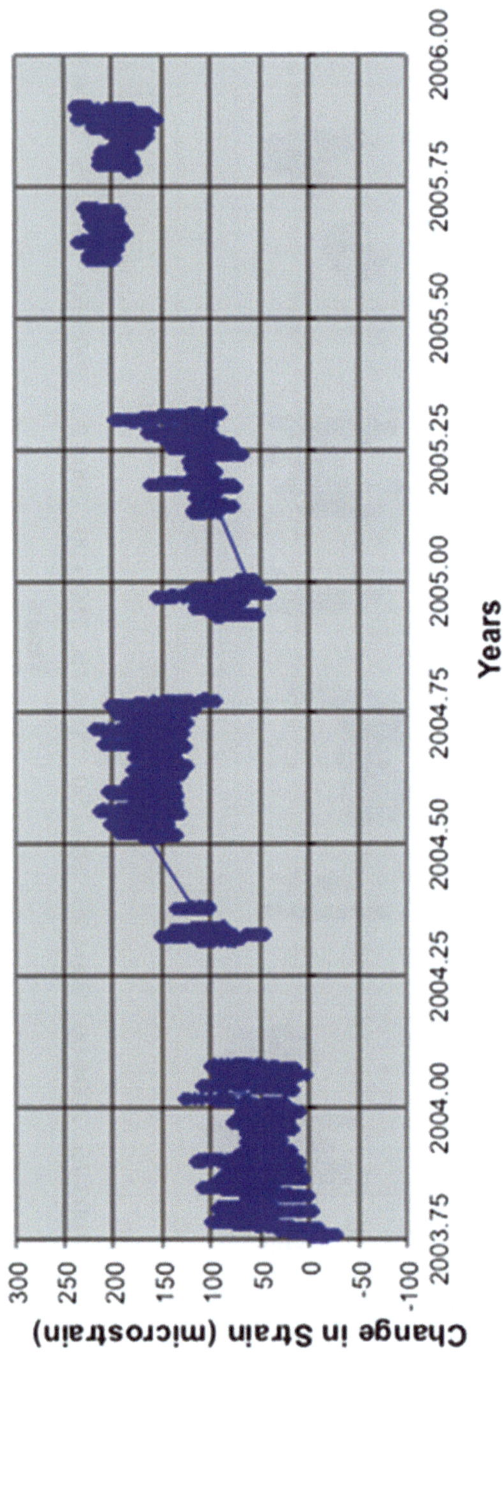

Fig. 11.5 Strain-time history of a sensor on the CFRP wrap, near the anchorage

11.8 Field Observations

The sensors' data were supplemented by a field inspection of the structure on July 10, 2006. The field investigation was prompted by unexpected sensor readings and the observation of cracking on the structure during maintenance of the SHM system batteries. The arms-length inspection was facilitated by the use of the Ohio Department of Transportation's (ODOT) snooper truck. Inspection consisted of sounding the suspect patching material to estimate the presence of voids and debonding. The CFRP wrap near the anchorages was also tapped to estimate the debonding of the CFRP wrap from the concrete and patching.

The inspection found that several of the repaired beams exhibited cracking at the interface between the patching material and the original sound concrete. Cracking appeared both in repaired beams that were strengthened, as well as in beams that were not strengthened. A segment of unsound and cracked patching was broken off for additional inspection, which revealed a white substance likely to be efflorescence, as seen in Figs 11.6 and 11.7. There is no indication of corrosion of the prestressing strands although some steel rebar corrosion was noted on the Clinton Street Bridge. Figure 11.8 displays the bottom of the beam and shows mapped cracking in the patching material with efflorescence leeching through the cracks. Figure 11.9 displays the side elevation of the beam and shows cracking at the interface between the patch material and the concrete substrate. Figure 11.10 shows the bottom of a beam strengthened with the CFRP stress head and a non-stressed CFRP plate. Detailed material properties are beyond the scope of this report. Relevant data can be found in Sika CarboDur's product data sheet (Sika 2002). Figure 11.11

Fig. 11.6 Removed section of unsound patching material, looking at the side of the beam

Fig. 11.7 Cracking observed during inspection, looking at the bottom of the beam

Fig. 11.8 Cracking observed at interface of patch and concrete beam

exhibits the installed sensors adjacent to the anchorage system. This image also shows the additional FRP reinforcement used surrounding the anchorage zone.

The occurrence of concrete cracking without steel corrosion in Fig. 11.6 directed us to investigate other causes. FHWA's Petrographic method of examining hardened concrete (FHWA 2004) suggests that one should look for three items when performing field observations that may indicate alkali-aggregate reactions (AAR): (1) map

Fig. 11.9 Cracking observed at interface of patch and concrete beam

Fig. 11.10 Sika stress head S624 strip and S1012 strip (Sika 2002)

cracking, (2) structural evidence of expansion, and (3) unusual pore fillings. Map cracking and white efflorescence, as seen in Fig. 11.7, are consistent with those described and shown in the FHWA manual. When considered with the expansive trends in the sensor data, this evidence leads to the conclusion that the cracking was primarily caused by an AAR.

Fig. 11.11 View of anchorage system including installed sensors and FRP reinforcement

AAR occurs when alkalis within the concrete or from outside sources such as deicing salts or saltwater react with siliceous or carbonate aggregate (TRB 2006). The most common type of AAR is alkali-silica reaction or ASR. The reaction occurs when a reactive aggregate, alkalis, and moisture that are present produce an alkali-silica gel which expands when water is absorbed. Petrographic or other analysis of the Hopkins Street and Clinton Street beams was not performed to verify the presence of these materials; however, ASR has been identified as a cause for deterioration in concrete in Ohio (FHWA 2004). Also, alkalinity and moisture from deicing salts were noted as a cause of the initial deterioration prior to repair and probably remained in the concrete following repair. The cementitious patching material also introduced more moisture and alkalinity.

A number of the sensors on both bridges exhibited expansive behavior as a result of the ASR effect. The expansive behaviors were not observed in great magnitude in the first 6–9 months following construction. This could be attributed to the reactions that created alkali-silica gel not having reached the critical level necessary to produce cracking. This time period also followed the first exposure to the wet spring season and more humid summer season, providing the moisture necessary for the reaction.

11.9 Full-Scale Live Load Tests

In addition to the visual inspections and long-term structural health monitoring, described above, a series of full-scale live load tests were conducted prior to rehabilitation and upon completion of strengthening systems. The intent of initial field tests was to determine a baseline load rating to be utilized for post-strengthening requirements and to provide a basis for comparison with the structures' performance following rehabilitation. The primary focus of this report is the structural-health monitoring application. Only a brief overview of the live-load test is presented here. Details of live-load test results are beyond the scope of this study and are included in comprehensive reports (BDI 2003).

A total of 52 strain transducers were utilized for instrumentation of the Hopkins Street Bridge, whereas, a total of 48 strain transducers were used to instrument the Clinton Street Bridge. BDI strain transducers ST-350 were employed for all measurements. The 3-in (76 mm) long strain transducers are rugged and can be installed expeditiously in any weather (BDI 2010). Other key features of the BDI transducers include easy installation (can be installed in 5 min or less), compatibility with most data acquisition systems, complete reusability, and cost-effectiveness (BDI 2010). The access to underside of bridge superstructure was made possible via a snooper truck, used routinely by the department of transportation crew for inspection purposes. A forklift, carrying concrete blocks, was employed to enable the live-load tests of Clinton Street Bridge's walkway beams. The exact weight of the loaded forklift was determined via a portable scale and the load test was repeated twice to ensure reproducibility and integrity of results and testing procedures. The bridge response was monitored in terms of strains while the vehicle's position along a prescribed path was recorded remotely.

A three-axle dump truck, with known axle loads, was used to perform the full-scale live-load tests of Hopkins Street Bridge. The test vehicle was driven along three predefined lateral paths, while strains were being measured during each truck crossing and the vehicle's position was monitored remotely. Similar to the Clinton Street Bridge scenario, each test was repeated twice along each prescribed path to ensure reproducibility of the test results and testing procedures. The view of the bridge underside, with several installed strain transducers, is exhibited in Fig. 11.12. A typical response history is presented in Fig. 11.13, representing the plot of load position versus micro-strain at the mid-span of a deteriorated beam.

11.10 Conclusions

The long-term health monitoring system via the vibrating wire sensors met the needs of the current project by providing stable measurements of strain and temperature. In addition, all the strain measurements from full-scale live load tests reveal linear and elastic response of both bridge structures with respect to load magnitude (truck position). Based on the long-term health monitoring results and live-load tests:

- The largest strains were identified at mid-span of the exterior beams, where the deterioration was greatest.
- Strain histories of the Clinton Street Bridge tests attest to the continuous behavior of beams under the applied live-loads. This conclusion confirms the actual pier connection details exhibited in design plans.

Fig. 11.12 Strain transducers attached to the underside of the bridge

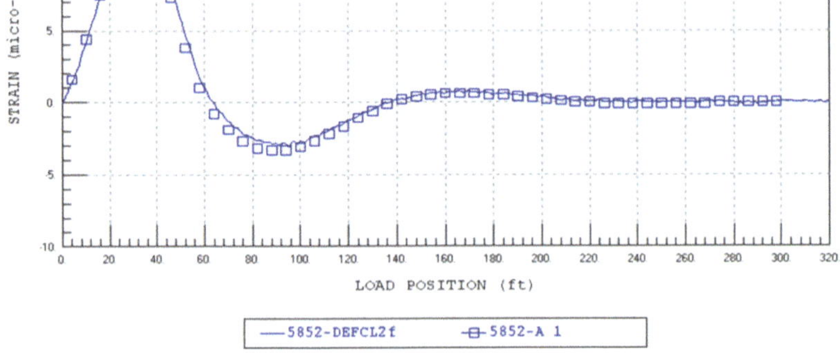

Fig. 11.13 Typical response history of a deteriorated fascia beam

- In the case of Clinton Street Bridge, the measured strains were relatively small. The maximum (live-load) strain was 17 micro-strain, corresponding to a stress intensity of 85 psi (assuming a modulus of 5,000 ksi for the prestressed concrete).
- The data collected exhibits evidence of expansive alkali-silica reactions and continued deterioration that was also visibly observed on the beams.
- Furthermore, the continued deterioration of the beams is affecting the level of stress in the CFRP Stress Head plate.
- The data also has shown that the levels of stress in the CFRP Stress Head plate have changed due to the anticipated losses. The losses appear to be less than assumed during the design process.
- The use of the stress-head post-tensioning system proved to be a viable and effective means to strengthen a deficient bridge.
- The level of cracking observed in the patching material and redistribution of stresses, seen in the sensor data and attributed to the cracking, indicate that the patching material is no longer adequately bonded to the beam.
- Based upon the data collected during post-tensioning, it is apparent that compression is induced in the bottom of the beams thereby improving the load carrying capacity.

11.11 Recommendations

The long-term health monitoring system performed well for the most part; however, some improvements for future similar applications are recommended. The use of 12V batteries did power the system, but they are recommended only for nearby, easily accessible locations. The dataloggers can also be programmed to send out alerts concerning anomalies related to the battery capacity. The adhesive used to mount the sensors also proved to be problematic. Proper application of any adhesives is recommended and mechanical anchorage is even more effective.

The data collected from the long-term health monitoring and the field observations indicate that the beams have continued to deteriorate following the repair. The data do not indicate that any of the Stress Head assemblies have lost their post-tensioning force. However, due to the level of cracking noted in the field and indicated by the data, the redundant epoxy bond to the concrete beam cannot be considered an effective means to transfer the post-tensioning force into the concrete beam. The post-tensioning force, therefore, must be transferred into the beams by the anchorage dowels at either end of the Stress Head system. The sensor data collected at sensors near the anchorage assemblies and on the CFRP wrap have also shown signs of deterioration. Only one of the beams, part of the Clinton Street bridge superstructure, experienced debonding of the CFRP wrap placed around the anchorage area. Additionally, the sensor data, collected on the CFRP Stress Head plates, revealed increases in the level of tension of the system. Continued inspection and monitoring are recommended, particularly in the areas of the anchorages, due to the continued deterioration and the increases in tension noted as caused by the expansion of the deteriorating concrete.

A more detailed investigation of the deteriorating concrete and patching is recommended. The proposed study should attempt to quantify if alkali-silica reactions and deterioration are expected to continue and to what extent. The relevant guidelines for the investigation are included in Chapter 10 of FHWA's *Petrographic Methods of Examining Hardened Concrete: A Petrographic Manual* (FHWA 2004). The tops of the interior beams have been protected from deicing salts by a waterproofing membrane installed as part of the rehabilitation; however, the presence of reactive aggregates may indicate the need for additional maintenance to ensure durability.

References

Bridge Diagnostics Inc (2004) Load-testing and load-rating report: Hopkins Street Bridge, Defiance. Final report submitted to defiance county engineer's office, Defiance, Sept 2004

Bridge Diagnostics Inc (2010) BDI strain transducer ST-350. http://www.bridgetest.com/products/BDI%20Strain%20Transducer.pdf. Accessed on 15 June 2010

FHWA (Federal Highway Administration (2004) Petrographic methods of examining hardened concrete: a petrographic manual. FHWA-HRT-04-150

Geokon (2002) Instructional manual model VK-4100/4150 vibrating wire strain gages. Revision G 09/02, Lebanon

Grace NF, Enomoto T, Sachidanandan S, Purayankara S (2006) Use of CFRP/CFCC reinforcement in prestressed concrete box beam bridges. ACI Struct J 103(1):123–132, Jan/Feb 2006

ISIS Canada (2001a) Guidelines for structural health monitoring, vol 2, Design Manual. ISIS Canada, Winnipeg

ISIS Canada (2001b) Installation, use and repair of fiber optic sensors, vol 1, Design Manual. ISIS Canada, Winnipeg

Kessler SS (2002) Piezoelectric-based in-situ damage detection of composite materials for SHM systems. PhD Dissertation, Massachusetts Institute of Technology

Shepherd R (1964) Strain measurement using vibrating-wire gages. J Exp Mech 4(8):244–248

Sika CarboDur (2002) Heavy-duty CFRP strengthening system, product data sheet, 8 p

Transportation Research Board (TRB) (2006) Control of cracking in concrete, Transportation Research Circular E-C107. TRB, Washington, DC

Zoghi M, Foster DC (2006) Post-strengthening prestressed concrete bridges via post-tensioned CFRP laminates. SAMPE J 42(2):24–30

Chapter 12
Life-Cycle Cost Comparison for Steel Reinforced Concrete and Fiber Reinforced Polymer Bridge Decks

Brent Kawahara, Hector Estrada, and Luke S. Lee

Abstract The main goal of this work is to determine the economic feasibility of Fiber Reinforced Polymer (FRP) as the primary material for construction of short span bridge decks. The analysis is based on a comparison of the lifecycle costs for a cast-in-place steel reinforced concrete bridge and FRP bridge deck over a 60-year time horizon; both have dimensions of 8.0 m (26.3 ft) long by 8.3 m (27.2 ft) wide. The total cost estimate using the FRP material ranges from $82 to $182 per square foot ($882–$1958 per sq meter) with a range average of $132 per square foot ($1,420 per sq meter). It is also shown that the breakeven construction cost of the FRP deck is $113 per square foot ($1216 per sq meter) for a discount rate of 2%. The main factors affecting the analysis are costs of material, traffic delays, and discount rate.

Keywords Composites • Bridge decks • Life cycle analysis

12.1 Introduction

Civil infrastructure provides the basic framework for a functioning society. It has directly and indirectly influenced our civilization with the sole purpose of raising our standards of living through increased global interaction. Transportation, potable water, and land development are just a few infrastructure categories that are

B. Kawahara (✉) • H. Estrada
Department of Civil Engineering, University of the Pacific,
3601 Pacific Ave., Stockton, CA 95211, USA
e-mail: B_kawahara@u.pacific.edu; hestrada@pacific.edu

L.S. Lee
School of Engineering and Computer Science, University of the Pacific,
3601 Pacific Ave., Stockton, CA 95211, USA
e-mail: llee4@pacific.edu

R. Jain and L. Lee (eds.), *Fiber Reinforced Polymer (FRP) Composites for Infrastructure Applications*, Strategies for Sustainability, DOI 10.1007/978-94-007-2357-3_12,

encompassed by the field of civil engineering. These infrastructure systems are under increasing strain because of rapid population growth, aging, and natural disasters, resulting in increased demands on performance and cost without consideration for the environmental and social impacts associated with construction and infrastructure management. For example, the fabrication of cement results in large quantities of CO_2 emissions (Kendall 2004) which amounted to 71.3 million metric tons in 2008 (van Oss 2010). To mitigate the effects of emissions, new materials and production methods are under investigation such as fiber reinforced polymers (FRP) and engineered cementitious composites.

To limit our carbon footprint, environmental policies have been adopted and are now enforced; particularly in all aspects of infrastructure construction. New research has led to innovative materials such as engineered cementitious composites and FRP which have higher specific properties (strength-to-weight and stiffness-to-weight ratios). These new materials greatly decrease the superstructure load, which leads to smaller structural elements at the sub- and super-structure levels. Such products reduce design costs, but are also subject to the restrictions of existing environmental policies. Research is needed in order to determine the feasibility of using new materials to construct a product and compare the costs and environmental impacts of constructing the same product with existing materials. This study analyzes the use of FRP as an alternative to conventional cast-in-place reinforced concrete for the construction of a bridge. The bridge deck is the primary focus for our analysis as it will incur the majority of the cost difference from the manufacturing of the materials through the end of life disposal – cradle-to-grave.

12.1.1 Environmental Impact of Concrete Production

Concrete is one of the most widely used materials for infrastructure construction throughout the world. The primary ingredients in concrete are cement, aggregate, sand and water. Also, because of the poor tensile strength of concrete, steel reinforcing bars (rebars) are incorporated into the concrete. Reinforced concrete is the most commonly used material in structural design; primarily because of is lower cost compared to other materials. Although concrete has a considerable impact on the construction industry today, its environmental impacts have become a significant liability. These effects are based on carbon dioxide (CO_2) emissions from the production of cement. The primary processes in cement production that generate CO_2 emissions are the combustion of fuels to heat the kiln and limestone calcination (Kendall 2004). Other environmental impacts occur during the harvesting of raw materials used in cement production; such as conversion of land use, loss of wildlife habitat, erosion and sedimentation, noise and dust, blasting, chemical spills, and contamination of groundwater and surface water (Chandler 2004).

The U.S. transportation highway system is comprised of 599,766 bridges with 152,324 of them unsuitable for current or projected traffic demands as of August 2000 (ASCE 2009). The cost of maintenance and repairing all of these deficient bridges is estimated at $140 billion (ASCE 2009). The material of choice for the majority of these bridge projects is reinforced concrete. The main cause of bridge deterioration is corrosion of the steel reinforcement in the concrete decks (Nelson 2005). The harsh environment (e.g. marine environments) that bridges are exposed to as well as the required use of deicing salts in some regions can cause corrosion of steel reinforcement. The concrete degradation mechanism is primarily driven by the byproducts of corrosion reactions, which causes expansion of the concrete around the reinforcement resulting in cracking. Once these cracks develop, the steel rebar becomes exposed to the environment and an accelerated rate of corrosion occurs (Nelson 2005). According to the Federal Highway Administration, the corrosion of reinforced concrete bridge decks has caused an estimated $8.3 billion per year for highway bridges alone (Nelson 2005). Instead of replacing entire reinforced concrete bridges, researchers are considering the use of alternative materials to repair or replace damaged bridge decks. Among these alternatives are FRP composites, high-performance concrete, high performance steel, new aluminum shapes, and engineered cementitious composites (ECC) (Ehlen 1999). Because of the environmental impacts posed by concrete production alternate solutions have been sought in order to minimize the carbon footprint of infrastructure construction. One alternative being considered, is the use of FRP composites instead of reinforced concrete for bridge decks.

12.2 FRP Background Information

FRP composites, are comprised of a combination of a polymer matrix such as polyester or epoxy and a reinforcement fiber such as glass, carbon, or aramid. The two constituent materials are combined to produce a composite material, the strength of which is much greater than the sum of the constituents' strengths (Nelson 2005). These materials are engineered to produce a desired set of properties for a particular function, whether it be used in a bridge deck or other structural component. It is estimated that $30 million has been spent by the federal government in developing composites for use in infrastructure (Mu et al. 2006). The use of FRP was not common until after World War II, where commercial use was first established (Telang et al. 2006). The use of FRP in civil infrastructure applications has only recently occurred more prominently over the past 15 years (Telang et al. 2006).

12.2.1 Advantages of FRP Composites

The advantages associated with FRP include: high specific properties (strength-to-weight and stiffness-to-weight ratios), corrosion resistance, use in seismic rehabilitation, and low-cost of erection/maintenance (Telang et al. 2006). A more detailed description of each of these advantages is described as follows:

- **High specific properties** are particularly important in bridge decks; FRP bridge decks weigh approximately 20% of the weight of an equivalent reinforced concrete deck. This allows a reduction in dead load requirements, which in turn reduces the strength requirements for the sub- and super-structural elements. For existing bridges, a replacement FRP deck with a high strength-to-weight ratio can increase the live-to-dead load ratio, potentially resulting in adjustments to any imposed bridge load restrictions.

- **Corrosion resistance** is critically important in bridge decks exposed to corrosive and deicing chemicals. Corrosion of steel reinforcement is the primary cause of deficiency in reinforced concrete infrastructure. FRP composites are being proven as durable materials in infrastructure applications and, in the case of carbon fiber, corrosion resistant. The use of FRP bridge decks may result in a longer service life and lower maintenance costs, all of which contribute to lower life-cycle costs (Nelson 2005).
- **Seismic protection** of bridges using FRP composites is indirectly related to bridge decks. FRP jackets are typically applied to concrete columns to enhance their confined strength; a smaller bridge deck generates smaller inertial lateral forces requiring smaller FRP retrofit systems. Placing jackets around concrete columns confines the concrete from spalling, preventing steel bars from compression buckling and reducing structural damage more effectively than using internal steel ties. FRP jackets are preferred to steel jackets because they are not susceptible to corrosion, are easy to install, do not change aesthetics, and do not affect the stiffness of the members.
- **Low-cost erection/maintenance** is important for highly trafficked bridges, where the indirect costs from traffic delays and lost productivity are estimated at 10 times the direct costs (Nelson 2005). FRP bridge decks can be installed in approximately half the time of an equivalent reinforced concrete bridge deck (Nelson 2005). The faster installation time results in a lower economic impact and in lower environmental impacts with fewer emissions from construction equipment and transportation.

12.2.2 Disadvantages of FRP Composites

The disadvantages associated with FRP include: high unit costs and low ductility. A more detailed description of each of these disadvantages is provided as follows:

- **Material unit cost** when compared to reinforced concrete is much higher. The range of costs found for concrete and FRP were \$25–\$73 and \$50–\$170, respectively. Although, there is potential cost savings over the extended service life of FRP, engineers are typically hesitant to change designs utilizing an unfamiliar material (Nelson 2005).
- **Low ductility** results in little to no warning of failure, which can be potentially catastrophic (Nelson 2005). Reinforced concrete allows some warning before failure because of the ductility of the reinforcing steel, which can be seen in routine inspections such as excessive deflections (Nelson 2005). Due to this danger, inspections are conducted every year as opposed to every 2 years for bridge decks of accepted material characteristics (Telang et al. 2006)

12.3 Environmental Impacts of Composite Production Processes

As noted earlier, the production of cement has a high impact on our environment, with nearly 77 metric tons of CO_2 emissions per year (Chandler 2004). To reduce the impact of these greenhouse gases on the environment, a number of alternatives are currently being investigated – one of which is FRP composites. A report by Anderson et al. (2004) outlines the production stages, manufacturing processes, and assesses a rating system for multiple composite systems, including a double curvature panel, a flat sandwich panel, and a complex molded component. Table 12.1 shows a weighting scale used to assess the composites based on the specific environmental issues affected in the UK.

The fabrication processes analyzed included hand lay-up, spray-up, vacuum-bag molding, resin infusion, resin transfer molding (RTM), autoclave molding, pultrusion, and compression molding. Different materials were tested using each fabrication process and assigned a rating of A through E, with an A rating representing the lowest environmental impact. Some of the more important issues tested were human toxicity, waste disposal, ecotoxicity, and remuneration. These results were based on a survey that was restricted to one region of the UK; therefore, the remuneration data is limited in scope. The remuneration data is based on a five scales from A to E, with an A rating indicating the highest remuneration rates for the employee. The best process for the double-curvature panel was the compression molding in which three out of four material compositions yielded A ratings. For the flat sandwich panel, the highest processes were vacuum bag molding, RTM, autoclave molding, and pultrusion. The complex molded component had the highest rating when using the compression molding process with three out of four material compositions having an A rating.

Table 12.1 Environmental issues relative weight scale (Anderson et al. 2004)

Environmental issue	Relative weight scale %
Climate change	38.0
Fossil fuel depletion	12.0
Ozone depletion	8.2
Human toxicity to air	7.0
Waste disposal	6.1
Water extraction	5.4
Acid deposition	5.1
Eutrophication	4.3
Ecotoxicity	4.0
Summer smog	3.8
Minerals extraction	3.5
Human toxicity to water	2.6

12.4 Review of Literature

12.4.1 Composite Bridge Decks

Life-cycle cost analyses of FRP bridges in literature are scarce and those that have been published are only preliminary estimates. A study by Ehlen (1999) evaluated the life-cycle costs of FRP bridge decks using computer simulations. Three types of FRP bridge decks were evaluated in this study:

- The SCRIMP FRP deck was comprised of a 9-layer top plate of stitch-bonded e-glass, a 5-layer e-glass trapezoidal inner sandwich, and a 13-layer bottom plate of stitched-bonded fiberglass. The resin used was a liquid vinyl ester and was applied using vacuum pressure. The completed FRP bridge deck weighed approximately one-sixth that of a comparable concrete bridge deck.
- The wood-core was composed of bamboo sections bonded together with resin. These sections were then covered on the top and bottom with fiberglass layers. Each of these layers was comprised of fiberglass fabric that had been soaked in a vinyl ester resin before application.
- The last bridge deck evaluated was a PP deck manufactured using FRP pultrusion. Planks were created by pulling a fiberglass strand and fabric wetted with vinyl ester through a heated die.

The life cycle analysis included the costs associated with construction, maintenance, and disposal of the bridge decks. Each bridge deck was evaluated using a cost analysis that incorporated all project-related costs for each material. Each of the three cases was evaluated against a base case, which was a conventional concrete bridge deck of equivalent strength in areas of high and low traffic. After the simulation, only the wood-core FRP bridge deck was deemed to be a feasible alternative to the concrete bridge deck for areas of high traffic. The total cost of the concrete bridge deck was estimated at $984,569, that of the wood-core deck at $891,209, that of the SCRIMP deck at $1,108,230, and that of the PP bridge deck at $1,355,223. For areas of low traffic, no FRP bridge deck alternatives resulted in sufficient user cost savings to justify their application.

Nelson (2005) studied the behavior of GFRP bridge decks for highway bridges. The chosen decks were manufactured using DuraSpan® produced by Martin Marietta Composites (2004). Two bridge decks were evaluated in this study. The first was the DuraSpan® 766, with a depth of 19.5 cm, surface width of 30.5 cm, and weight of 92.8 kg/m^2. The second was the DuraSpan® 500 with a depth of 12.7 cm, surface width of 61.0 cm, and weight of 63.5 kg/m^2. Each of these designs was comprised of structural components bonded together to form panels that can be placed directly on top of a bridge base. These decks were comprised of E-glass fibers held together by a polyester matrix and manufactured using pultrusion. Various limit states were examined for each bridge deck to determine the load-deformation relationship and failure modes of the GFRP decks. The systems were tested for flexural capacity, coefficient of thermal expansion, and connection details.

From this information, a finite element analysis was performed for the bridge decks to predict and verify the behavior of each system. From the load tests the following conclusions were drawn: (1) The DuraSpan® bridge deck is a feasible alternative to conventional concrete bridge decks; (2) The ultimate strength of the DuraSpan® bridge deck is largely influenced by the web shear capacity and the interface between web and flange; (3) The material properties that had the largest influence on the proposed model's ultimate strength were the flange longitudinal modulus and the web shear modulus.

Daniel (2003) performed a life cycle analysis for a pedestrian bridge in Noordland inner harbor in the Netherlands. This bridge spanned 13.5 m with two free beams and a width of 1.6 m. The bridge was designed for a service load capacity of 4.0 kN/m^2 and a maximum allowed deflection of 1/250 of the beam span. Daniel chose to investigate five alternatives; structural steel, stainless steel, FRP composites, aluminum, and reinforced concrete. The criteria included in this analysis were construction costs, maintenance costs, service life, and environmental impact. The five different material alternatives were investigated in terms of energy consumption and environmental pollutants. Table 12.2 lists a cost estimate breakdown for each material design; of particular interest are the cost estimates for materials and maintenance for composites and conventional concrete, both approximately double for composites compared to concrete. Although costs of materials and maintenance for composites are higher, the amount of energy consumed is only half for composites compared to concrete. The "exergy" method was used to quantify the energy impact of each of the five materials evaluated. This method accounts for the potential energy within a material that allows it to perform work. Figure 12.1 shows the amount of energy consumption for each of the five materials at delivery and maintenance stages. Figures 12.2 and 12.3 show that a conventional concrete bridge yields nearly four times the amount of water and air pollutants compared to a composite bridge. This data was based on an eco-analysis that follows the critical load method of Mahadvi and Ries (1998).

Shimomura et al. (2009), examined the life cycle cost (LCC) and conducted a life cycle assessment (LCA) of an FRP bridge in Okinawa, Japan. This bridge was subjected to harsh environmental conditions near the seashore, which mainly affected the rate of steel corrosion that occurred within the bridge deck. Five different bridges were analyzed, two FRP and three pre-stressed concrete (PC). The FRP bridges were two-span girder type with GFRP C-girders spanning 36.9 m, with a width of 4.3 m. The PC bridges were single-span girder type with hollow post-tension concrete beams spanning 35.0 m with a width of 4.3 m. The FRP bridge decks are comprised of glass fibers held together with a vinylester resin, assembled by a combination of hand-lay up and pultrusion. The LCC study included costs from initial construction and maintenance, and neglected costs from demolition and discount rates. Figure 12.4 depicts a flow chart illustrating the life cycle process. The LCA was based on greenhouse gas emissions shown in Table 12.3, particularly those generated during material fabrication and project construction stages. The total LCC of all five bridges is summarized in Table 12.4. Cases 1–3 represent pre-stressed concrete bridges (Case 1 has no corrosion protection). Cases 4 represents a

Table 12.2 Cost breakdown for each bridge deck material design alternative (Daniel 2003)

	Criterion			
Materials	Initial costs (euros)	Maintenance costs (euros)	Environment: energy consumption	Environment: critical volume of polluted…
Structural steel	Painted: € 40.000	Painted: € 30.000	"exergy" method: 294,000 MJ	…water: 697.4 m^3
	Aluminum coated: € 50.000	Aluminum coated: € 6.000		…air: 7.09×10^6 m^3
Stainless steel	In steel AISI 316 L: € 110.000	In steel AISI 316 L: € 6.000	"exergy" method: 329,600 MJ	Not investigated but certainly more water and air pollution than for structural steel
	In steel AISI 304 L: € 96.000	In steel AISI 304 L: life cycle shorter		
Composites	Pultruded sections of fiberglass reinforced polyester (FGRP): € 70.000	Rough estimation: € 17.000	"exergy" method: 120,000 MJ	…water: 85.8 m^3 …air: 7.72×10^6 m^3
Aluminum	Quality AlMgSi1 acc. to DIN 1748: € 77.000	Rough estimation: € 19.000	"exergy" method: 268,700 MJ	…water: 565.3 m^3 …air: 41.10×10^6 m^3
Concrete	Concrete B35 reinforced, handrails etc. € 30.000	Rough estimation: € 10.000	"exergy" method: 277,200 MJ	…water: 341.9 m^3 …air: 31.04×10^6 m^3

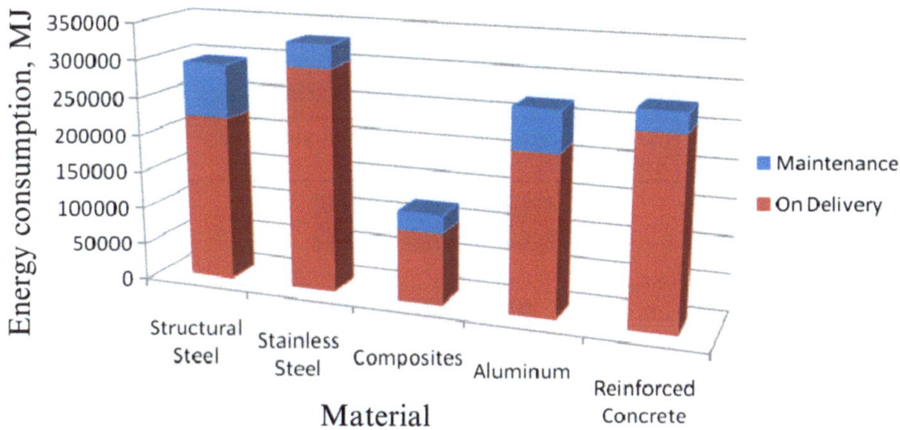

Fig. 12.1 Energy consumption of five bridge projects (Daniel 2003)

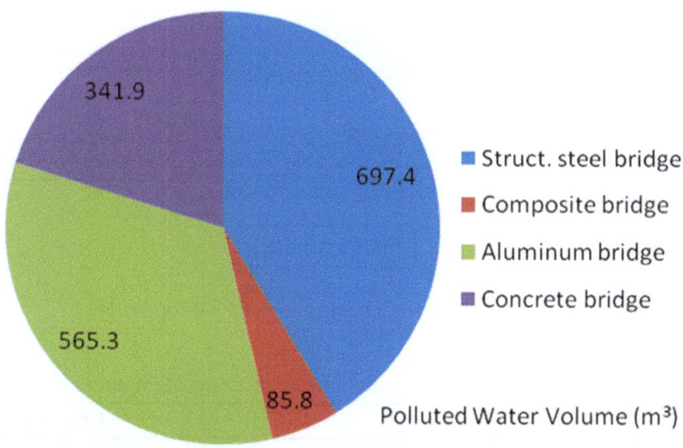

Fig. 12.2 Critical polluted water volumes comparison (Daniel 2003)

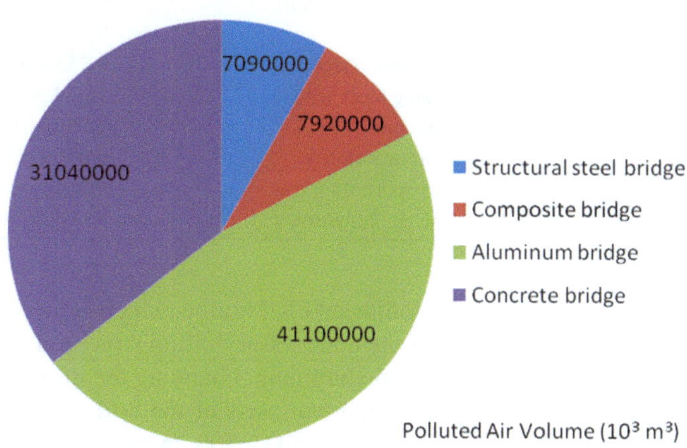

Fig. 12.3 Critical polluted air volumes in total comparison (Daniel 2003)

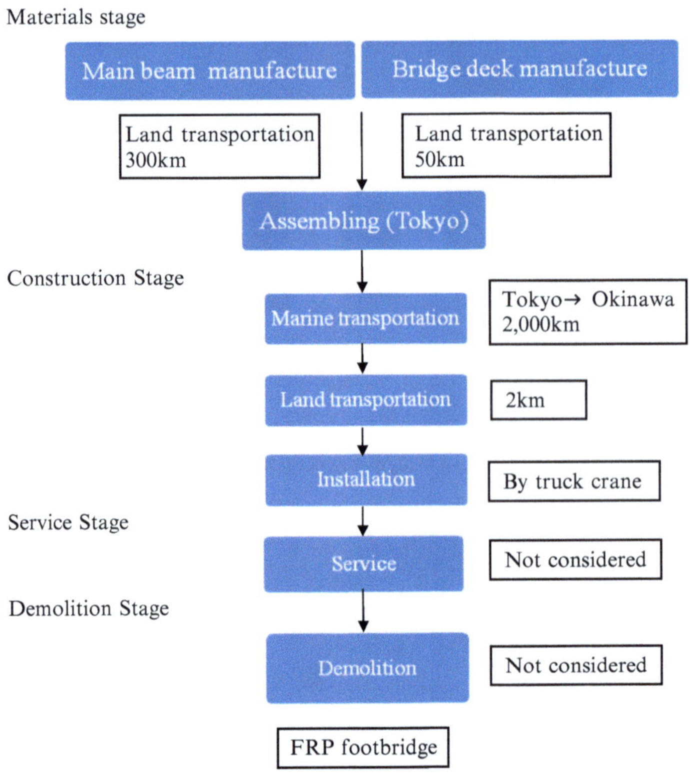

Fig. 12.4 Evaluation scheme of life cycle CO_2 emission of FRP footbridge

Table 12.3 Unit of carbon dioxide (CO_2) emissions by process (Shimomura et al. 2009)

	Process	CO_2 emissions
Materials	FRP (Hand lay-up)	4.97 $kgCO_2/kg$
	FRP (Pultruded)	3.09 $kgCO_2/kg$
	Concrete (F_c 27 N/mm^2)	0.918 $kgCO_2/kg$
		211.1 $kgCO_2/m^3$
	Reinforcement steel	0.755 $kgCO_2/kg$
	Prestressing steel wire	1.31 $kgCO_2/kg$
	Steel pipe pile	1.25 $kgCO_2/kg$
Transportation	Marine transportation	0.039 $kgCO_2$/tonne*km
	Land transportation	0.154 $kgCO_2$/tonne*km
Construction	Concrete (F_c 27 N/mm^2)	39.0 $kgCO_2/m^3$

standard FRP bridge and Case 5 an adjusted FRP bridge, which is the same as Case 4, but modified to decrease some of the unnecessary initial structural costs; including changing the handrail from FRP to aluminum, changing the design in the main girder joint, and sharing the mold for the two parts of the bridge. These revisions lowered the initial costs by approximately $112,500. Although the pre-stressed

Table 12.4 Total LCC of PC and FRP bridges (Shimomura et al. 2009)

	Case-1	Case-2	Case-3 Repair interval (RI) 15 year RI	Case-3 Repair interval (RI) 30 year RI	Case-4	Case-5
Superstructures initial cost	48,240	50,620	54,370		73,600	62,350
Substructures initial cost	10,130	10,130	10,130		6,910	6,910
Total initial costs	58,370	60,750	64,500		80,510	69,260
30-year maintenance cost	24,500	0	18,000	9,000	6,000	3,500
50-year maintenance cost	42,500	0	27,000	9,000	10,000	3,500
100-year maintenance cost	69,500	24,500	54,000	27,000	20,000	7,000
50 years LCC	100,870	60,750	91,500	83,500	90,510	72,760
100 years LCC	127,870	85,250	118,500	91,500	100,510	76,260

concrete Case 2 was the lowest for the 50 year LCC, the FRP Case 5 had a lower LCC over 100 years because of the lower maintenance costs. It is interesting to note that the post tensioned concrete limits the amount of cracking in the deck, allowing more resistance to corrosion than conventional concrete. A conventional concrete deck subjected to these harsh environmental conditions would require more maintenance than a pre-stressed deck, and likely replacement of the whole structure within the life cycle. The higher costs associated with the maintenance of a conventional concrete deck would surpass any initial savings, making the LCC higher for conventional concrete than any other alternative.

Nystrom et al. (2003) studied the financial viability of short span FRP bridges using the "learning curve" approach (also known as economies of scale); they concentrated on bridges of less than 10 m in length. This type of analytical procedure has been motivated by a 460% increase in composite applications from 1970 to 2000. The "learning curve" theory predicts that as the production of composite products increases, costs will generally decrease exponentially. For example, a product having an 80% learning curve will decrease in cost by 20% every time the production quantity doubles. Expenses included in the analysis are construction, life-cycle, future estimates, and indirect costs, all of which were divided into two categories: processes and materials. Under processes were manufacturing, quality control, manufacturability design, assembly design, transportation, assembly methods, and equipment. The materials were the equipment to manufacture, transport, and assemble the bridges.

Table 12.5 summarizes costs associated with a composite bridge built on the campus of University of Missouri-Rolla, the St. James Bridge located in Missouri, and a future bridge with similar costs to those of the St. James Bridge. This future bridge's cost estimate is based on the assumptions that 1,000 similar bridges (at a rate of 100 per year) have been constructed. The total cost per square meter for this future bridge is $740. The total direct costs for a comparable RC bridge would be $433 per square meter. Based on initial costs alone it is clear that an FRP bridge deck would not be cost effective compared to a reinforced

Table 12.5 Estimated costs for FRP bridges ($/m²) (Nystrom et al. 2003)

Cost categories	UMR	St. James (1)	Future (2)	% Change from 1 to 2
Size (m²)	26	67	67	
Material				
Fiber	$581	$183	$146	–20
Resin	$171	$134	$107	–20
Pultrusion	$1,136	$150	$113	–25
Others	$536	$19	$18	–5
Total Materials	**$2,424**	**$486**	**$384**	**–21**
Handling		$32	$19	–41
Manufacturing labor	$349	$102	$41	–60
Manufacturing overhead	$118	$177	$106	–40
Total factory costs	**$2,891**	**$797**	**$550**	**–31**
Shipping	$33	$15	$14	–10
Installation				
Labor	$19[a]	$19	$11	–40
Equipment	$19[a]	$19	$11	–40
Miscellaneous materials	$11[a]	$11	$7	–38
Total installation costs	**$49[a]**	**$49**	**$29**	**–39**
Total Installed Costs	**$2,973**	**$861**	**$593**	**–31**
Overhead rate		33%	25%	
Overhead costs	$284[a]	$284	$148	–48
Total direct costs (rounded off)	**$3,250**	**$1,150**	**$740**	**–35**

[a]St. James bridge costs were used to allow for comparison

Table 12.6 Life-cycle costs (Nystrom et al. 2003)

LCC description	RC bridges	Future bridge	FRP premium
Construction costs	$430/m²	$740/m²	$310/m²
Impact of disposal	$45/m²	$9/m²	($36/m²)
Impact of replacements	$119/m²	$67/m²	($52/m²)
Total cost including disposal and longer life	$594/m²	$816/m²	$222/m²

concrete bridge deck. A life-cycle cost approach was then used to evaluate the two bridges over the expected life, including the cost of disposal. Table 12.6 summarizes the evaluation of the RC bridge and future FRP bridge. The table also tabulates the cost difference between the two alternatives, which has a premium of $222 per square meter for the FPR bridge deck. This analysis however did not include maintenance, environmental, or social costs.

Unfortunately, a true comparison amongst all the cases presented in this section can not be made because the analyses are based on a number of assumptions and not all costs were accounted for in the final conclusions. Table 12.7 presents a summary of results of the studies presented in this section. From Table 12.7, it is clear that Ehlen has the most complete study; however, his analysis did not include environmental costs.

Table 12.7 Literature review summary

Author	Product/Item/ Component	LCC completed	Categories considered								
			Materials	Transportation	Construction	Maintenance	Inspection	End Life	Learning Curve	Social	Environmental
Ehlen	GFRP deck (SCRIMP) GFRP deck (WC) GFRP deck (pultruded)	x	x	x	x	x	x	x	x	x	–
Daniel	GFRP deck (pultruded)	x	x	x	x	x	x	–	–	x	x
Nelson	GFRP deck (pultruded)	–	–	–	–	–	–	–	–	–	–
Nystrom	GFRP deck (Honeycomb sandwich panels)	x	x	x	x	–	x	x	x	–	–
Shimomura	GFRP deck (hand-lay up & pultruded)	x	x	x	x	x	x	–	–	–	–

12.4.2 Life-Cycle Analysis of Bridges

A study completed by Chandler (2004) for the Center for Sustainable Systems at the University of Michigan, compared life-cycle cost models for conventional concrete (CC) and engineered cementitious composites (ECC). The study analyzed a collection of agency and social costs stemming from manufacturing of resources through material disposal. The agency costs were classified as expenses charged to government agencies, and the social costs included all external and other costs not included in direct costs. The life-cycle analysis study included a Life Cycle Inventory model and a Traffic-Flow model. The input data for the model came from Ecobalance Database for Environmental Analysis and Management (DEAM) database. Following are the five modules incorporated into this life cycle analysis: emissions-related costs from construction activities, emissions-related costs from construction traffic congestion, construction-related traffic crashes, extra vehicle operating costs, and user delay costs. The emissions-related costs included costs associated with fossil fuels used to distribute and transport construction materials, and combustion of machinery used to produce the raw materials. A standard discount rate of 4% was applied for construction projects per federal government recommendations (Chandler 2004).

Five traffic scenarios were defined and tested, which included each of the above module parameters. The variables tested were the percentage annual average daily traffic (AADT) growth and the percentage daily traffic using detours; both were evaluated at 0%, 5% and 10%. Five different scenarios were tested:

- Scenario 1 is the base case where both variables were kept at 0.
- Scenario 2 tested detour rate of 5%, with an AADT growth of 0%.
- Scenario 3 tested detour rate of 10%, with an AADT growth of 0%.
- Scenario 4 tested detour rate of 0%, with an AADT growth of 1%.
- Scenario 5 tested detour rate of 0%, with an AADT growth of 2%.

Chandler concluded that ECC outperforms CC across all life cycle cost stages as well as in the overall life-cycle cost. Also, the discount rate applied had no effect on the outcome of the lower life cycle cost. This was because environmental and user costs were lower for the ECC system by approximately 12%. The use phase and social costs were the critical factors in the results of the life-cycle cost analysis. The analysis concluded that AADT is the most important variable, and least important was the detour percentages.

A study performed by Kendall in 2004 also evaluated the use of ECC as an alternative to CC bridge decks. The parameters taken into account were materials, construction, traffic, distribution, and end-of-life. Supporting data was generated with the following programs: US EPA's MOBILE6.2 and NONROAD emissions models, and the KyUCP traffic model. The deck designs were tested over a 60-year time period with multiple variables such as annual growth rate and detoured traffic percentage. The majority of the changes in results were linked to the change in construction related traffic. Key findings included lower environmental impacts as well as lower energy consumption for ECC when compared to CC. One way of measuring the environmental effects is by greenhouse gas emissions. The ECC bridge deck showed a 33% reduction in greenhouse gas emissions when compared to the CC system. The primary contributor to these results was CO_2, which represented 98% and 99% of the emissions for the ECC and CC bridge decks, respectively.

12.5 Life-Cycle Cost Analysis

In this section we present life-cycle cost analyzes to evaluate the economic feasibility of a FRP bridge deck compared to a Steel Reinforced Concrete (SRC) bridge deck. This study is focused on short-span bridges with a total span less than 10 m, which comprise over half of all bridges in the United States (Nystrom et al. 2003). The bridge specifications for this study are similar to those of the St. James Bridge discussed in Nystrom (2003). The dimensions are 8.0 m (26.25 ft) long by 8.33 m (27.33 ft) wide. The FRP bridge deck consists of four self-supporting FRP honeycomb sandwich panels, each 600 mm (23.6 in.) thick. Also, since the spans of the two systems are equal, we assume that the supports and foundations are the same size for the two alternatives. However, given that the FRP composite deck will be lighter, the required foundation for this system could be made smaller, at a lower cost that for the SRC deck.

The life expectancy of the FRP bridge deck is assumed to be 60 years (Nystrom et al. 2003), while that for the SRC bridge deck is assumed to be 30 years (Chandler 2004). Therefore, two bridge deck replacements will be required for the SRC cost estimate over the 60-year time horizon of the FRP bridge deck. To compare the costs of the two cases, a life cycle of 60 years is assumed; that is, the SRC bridge deck will be rebuilt after 30 years. The total costs of bridges over their lifecycle include the initial maintenance (including preventive maintenance), inspection, repair, and non-agency (social) costs. To perform a lifecycle cost analysis, the time value of money must be taken into account. Also, to be able to compare the two systems on the basis of a lifecycle, costs must be discounted to present values using the expression:

$$C_{PV} = C_{CP} / (1 + r)^n$$

where, C_{PV} is the cost discounted to present value, C_{CP} is cost at current price, n is the time period in years, $1/(1+r)^n$, or $(1+r)^{-n}$ is known as the discount factor, and r is the discount rate (Biondini and Frangopol 2008). For this study, the discount rate is defined as the difference between the rate of return on a risk-free investment (such as the interest rate on 1 year Treasury bills) and the inflation rate (Chen and Lui 2005). An analysis provided in Chen and Lui (2005) uses historical discount rates from years 1962 to 2002, with an average discount rate of 2.07%. For our analysis, we will assume a discount rate of 2%. Also, details of a conventional concrete bridge analysis are presented first, followed by the FRP deck lifecycle cost analysis.

Table 12.8 Overview of construction activities

Construction activity	Frequency	Description	Dimensions
Deck replacement	Every 30 years	Replacing the full depth of the deck and re-pouring new concrete	Length: 8 m Width: 8.3 m Depth: 10 in
Deck resurfacing	Every 15 years	Removing top 2 in of concrete along the entire deck and re-pouring new concrete	Length: 8 m Width: 8.3 m Depth: 2 in
Deck maintenance	Every 5 years following a deck resurfacing	Chipping out a square/rectangle shape and re-pouring with concrete	Depth: 2 in

12.5.1 *Steel Reinforced Concrete Bridge Deck Cost Estimate*

The life-cycle cost analysis presented here is based on costs from materials, transportation, construction (labor, equipment, fuel, and miscellaneous costs), maintenance, end of life disposal (labor, equipment, fuel, and miscellaneous costs), environmental (primarily CO_2), and social costs (traffic congestion, delays, and accidents). In this study, we focus on the three bridge deck construction activities listed in Table 12.8. Deck replacement is by far the most expensive and the most material- and time-intensive construction activity. Deck replacement entails rebuilding the concrete superstructure; deck resurfacing entails replacing the top two inches of concrete; and deck maintenance entails fixing potholes and other routine maintenance activities. The lifecycle cost analysis presented here will cover each stage of the life of a bridge deck, from manufacturing to end-of-life disposal (Fig. 12.5)

12.5.1.1 Installation and Replacement Costs

These costs can be obtained from available data of previous construction projects, or by a detailed analysis of the costs associated with each stage of the initial construction/replacement processes. The California Department of Transportation publishes construction statistics (Caltrans 2010) and average highway contract prices (Caltrans 2009) for the State of California – one of the states with the largest number of bridges. Based on the construction statistics data, the average price per square foot of a reinforced concrete bridge deck was $247 in 2009. This same report estimates that the cost in 2010 for the same type of bridge will be between $90 and $200 per square foot for common span ranges of 16–44 ft, which we can use to interpolate a cost of $130 per square foot of the proposed 26.25-foot span bridge deck. Detailed cost analysis, materials (concrete and steel), and construction (labor and fuel) are included for the initial and replacement cost calculations.

Deck materials: Chandler (2004) provides the cost of concrete at $59/cyd (59 U.S. dollars per cubic yard) or $77/m^3. A more current estimate of $160/cyd is provided by Caltrans (2010). An additional cost of $40/cyd should be added to the

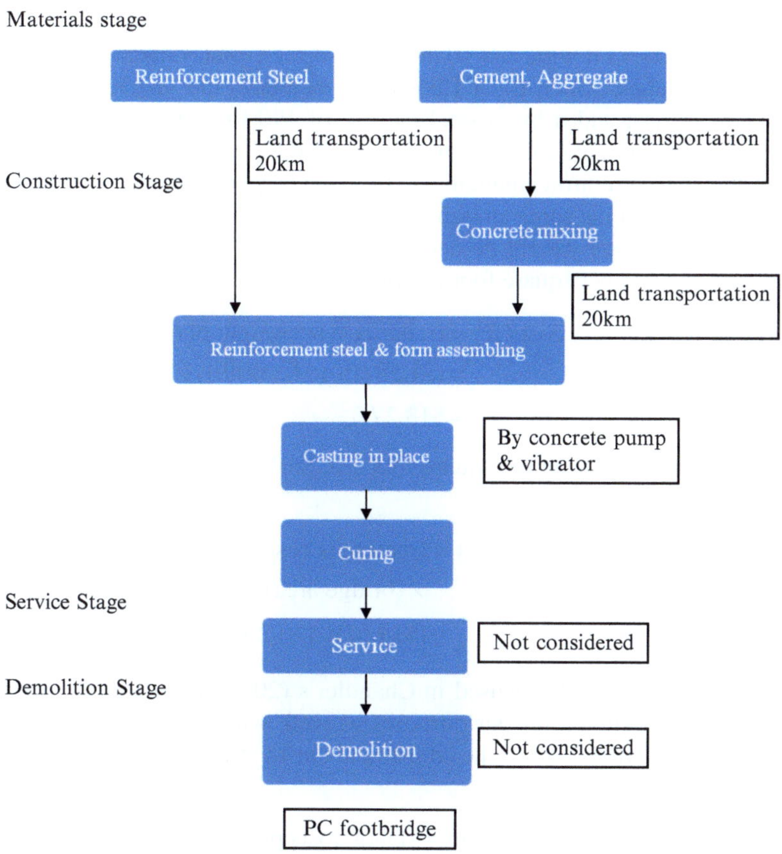

Fig. 12.5 Evaluation scheme of life cycle CO_2 emission of PC footbridge

materials for mixing and delivery (Chandler 2004); this is based on 2004 costs and should be adjusted for 2010 costs as follows: assuming an inflation rate, i, of 3% and a 6-year period, n, and using the present value of money equation, \$40/cyd $(1+i)^n=$ \$40/cyd $(1+0.03)^6=$ \$48/cyd. This results in approximately \$208/cyd (or \$7.7/ft^3) for the concrete material alone. An estimate of the cost for concrete per square foot, assuming a slab thickness of 10 in, is:

$$\begin{aligned}\text{Deck concrete replacement cost / ft}^2 &= (\text{cost per cubic foot})\\ &\quad\times(\text{depth of deck replaced})\\ &= (\$7.70/\text{ft}^3)\times(10/12\text{ft})\$6.4/\text{ft}^2\end{aligned}$$

The cost of steel reinforcing bar (rebar) is indirectly included in Chandler's analysis, but will be explicitly estimated in our life cycle cost analysis. The cost of rebar is estimated at \$1.58 per pound by Caltrans (2010). The mass of steel rebar needed for the proposed bridge deck can be estimated from a typical reinforcement ratio (i.e., cross sectional area of steel to concrete ratio) of 1.0% for bridge decks. For the proposed deck (27.33 ft wide and 10 in deep) we get a total concrete section area of 3,280 square inches, which for a 1.0% reinforcement ratio yields a steel area of 32.8 square inches; and using a steel density of 490 lbs per cubic feet, we get a requirement of 1.5 ton of rebar for our project. Also, since the deck is approximately square and we need reinforcement in the direction perpendicular to the traffic direction, we assume that the same amount of steel is needed in the perpendicular direction, 1.5 ton. The total weight of steel, 6 ton or 6,000 lb, can be used to find the current cost of rebar per square foot of bridge:

$$\begin{aligned}\text{Deck rebar replacement cost} / \text{ft}^2 &= (\text{rebar cost})\times(\text{weight of rebar}) / (\text{bridge area})\\ &= (\$1.58/\text{lb})\times(6{,}000\text{lb}) / (26.25\text{ft}\times 27.33\text{ft})\\ &= \$13.2/\text{ft}^2\end{aligned}$$

The total materials costs per square foot are: $\$6.4/\text{ft}^2 + \$13.2/\text{ft}^2 = \$19.6/\ \text{ft}^2$. The total materials cost is:

$$\begin{aligned}\text{Material cost of deck replacement} &= (\text{materials costs per square foot})\\ &\quad\times(\text{bridge area})\\ &= (\$19.6/\text{ft}^2)\times(26.25\text{ft}\times 27.33\text{ft}) = \underline{\$14{,}061}\end{aligned}$$

Deck labor: the cost of labor used in Chandler's (2004) study was $9 per square foot and will be assumed for our purposes as well, which is reasonable given the advances in productivity; that is, no adjustment due to inflation is required. The total labor cost is:

$$\begin{aligned}\text{Labor cost of deck replacement} &= (\text{labor costs per square foot})\\ &\quad\times(\text{bridge area})\\ &= (\$9/\text{ft}^2)\times(26.25\text{ft}\times 27.33\text{ft}) = \underline{\$6{,}457}\end{aligned}$$

Deck fuel and equipment rental: construction machinery fuel consumption is accounted for in terms of fuel consumed and emissions released. Table 12.9 lists general equipment needed for construction of a conventional concrete bridge deck (Kendall 2004). In this study, we assume a fuel cost of $3 per gallon for both gas and diesel for the West Coast (EIA 2010). A total fuel cost of $40,143 was used for the deck replacement phase of the bridge construction conducted by Kendall (2004). Dividing by the total square footage for that project yields a cost of $1.88 per square foot for fuel cost. This value is similar to the cost of fuel consumption derived in Chandler's study, $2.06 per square foot. The average of these two values is approximately $2.00 per square foot, which we will assume for our study. We also assume the cost of equipment rental to be 25% of the cost of fuel; therefore, we use $1.25\times \$2/\text{ft}^2 = 2.50/\text{ft}^2$ for fuel and equipment rental costs.

$$\begin{aligned}\text{Fuel cost of deck replacement} &= (\text{fuel costs per square foot})\\ &\quad\times(\text{bridge area})\\ &= (\$2.50/\text{ft}^2)\times(26.25\text{ft}\times 27.33\text{ft}) = \underline{\$1{,}794}\end{aligned}$$

Table 12.9 Deck installation/replacement fuel consumption (Kendall 2004)

Equipment	hp rating	Hours used	GMPH[a]	Total gallons
Crawler-mounted hydraulic excavator	428	128	12.36	1,582
Air compressor	350	64	14.56	932
Concrete mixer	8	0	0.23	0
Concrete paving machine	250	96	7.22	693
Concrete truck	300	32	8.66	277
Crane, 50 ton	177	176	5.11	899
Dumper	23	128	0.66	85
Hydraulic hammer	100	64	2.89	185
Motor grader	165	0	4.76	0
Signal boards	6	18,000	0.25	4,494
Vacuum truck	177	0	5.11	0
Water truck	450	0	12.99	0
Wheeled front-end loader	235	624	6.78	4,234
Fuel cost ($3 per gallon)				$40,143

[a]Gallons used per machine hour

Deck demolition: one of the end-of-life disposal costs is demolition, including labor and fuel. These costs are taken into account for a 60-year horizon. For this study, the SRC deck would need to be replaced once, which results in two bridge deck disposals. Equipment for demolition of a SRC deck can vary, with the most common options being hydrodemolition and conventional demolition. Hydrodemolition is a process where high-velocity water jets are used to remove or demolish concrete (Rampart Hydro Services, http://www.rampart-hydro.com/hydrodemolition.htm). This method is used in this study because it is faster than conventional methods and eliminates potential damage caused by vibrations. First, we consider the fuel cost, which is determined based on a fuel consumption of 30,799 L in Chandler's study for the removal of 832,931 kg of concrete. Utilizing the fuel-to-concrete mass ratio from Chandler (2004), the volume of fuel needed for the proposed span bridge deck can be determined. The amount of concrete removal for our bridge deck is only 4.61% that of Chandler's (2004) study, resulting in 1,421 L or 375 gallons. With the current gas/diesel price of $3 per gallon this amounts to approximately $1,126. However, demolition will not occur until 30 years from now; therefore, we must discount this amount to current value by using the discount rate factor, $\$1{,}126/(1+r)^n = \$1{,}126/(1+0.02)^{30} = \$622$, assuming $n = 30$ years and a discount rate, r of 2%. Also, in this case, labor is directly linked to equipment use; therefore, labor and equipment costs should be proportional to fuel costs, and we assume that equipment rental and labor costs are equal to the cost of fuel, $622. The total demolition cost is $1,244.

Deck disposal: all the demolition materials (concrete and steel) can be recycled. However, the most cost effective option for concrete is to transport it to the landfill.

Concrete recycling facilities will accept the materials at a rate of approximately \$12 per ton (Roxsand 2007). This results in a savings of approximately 67%. However, before the concrete can be recycled it must be ground into smaller pieces, which would require the material to be transported to a crushing center or rent portable recycling machinery. This results in longer construction times and traffic delays, raising the demolition costs higher than the cost of diverting the material to the landfill. The median landfill tipping fee cost is \$36.5/ton for California as given by CalRecycle (2010). We use this value for disposal of demolition materials. The mass of concrete removed from the bridge deck can be calculated by taking the volume of concrete needed for the construction activity (27.33 ft × 26.25 ft × 0.83 ft) and multiply it by the density of dry concrete (150 lb/ft^3), which is 89,677 lb or 44.8 ton. The cost of landfill disposal is:

$$\begin{aligned}\text{Landfill tipping costs} &= (\text{mass of wasteconcrete}) \times (\text{fee / ton}) \\ &= (44.8 \text{ tons}) \times (\$36.50 / \text{ton}) = \$1{,}637\end{aligned}$$

The assumed weight capacity of a truck to transport the concrete is 24 ton (Chandler 2004). Therefore, we need a total of two truck loads to transport the demolished concrete to the landfill. According to Bradley (2000), dump trucks are classified as class 8 vehicles. Vehicles in this class have a fuel efficiency of 6.6 mpg. Assuming a distance of 20 miles to the disposal site, and a cost of fuel of \$3/gallon, we get the transportation cost as follows:

$$\begin{aligned}\text{Transportation fuel costs for concrete demolition} &= (\text{truck loads}) \times (\text{distance to landfill}) / (\text{fuel efficiency}) \times (\text{cost of fuel}) \\ &= (2 \text{ trucks}) \times (20 \text{ miles}) / (6.6 \text{ mpg}) \times (\$3.00 / \text{gal.}) \\ &= \$18\end{aligned}$$

In this case, labor is also directly linked to equipment use, labor and equipment costs should be proportional to fuel costs; we will assume that equipment rental and labor costs are equal to the cost of fuel. That is, \$18.

The steel reinforcement in the concrete deck can be recycled. The steel can be removed from the concrete for a negligible fee. A net profit of \$114 per ton of steel can be assumed (Chandler 2004), which adjusted for inflations, $i=3\%$, over 6 years is \$114/ton $(1+0.03)^6=\$136$ per ton. The total revenue from the recycled steel is represented as negative in our model because it's revenue and not a cost to the construction company.

$$\begin{aligned}\text{Recycled steel revenue} &= (\text{mass of steel}) \times (\text{price / ton}) \\ &= (3 \text{ ton}) \times (\$114 / \text{ton}) = (\$342)\end{aligned}$$

The total cost for disposal adjusted for a payout in 30 years is given by the sum of the costs of landfill use, fuel, labor, and recycled steel revenue adjusted using $n=30$ years and a discount rate, i of 2%.; that is, $(\$1{,}637+\$18+\$18-\$342)/(1+0.02)^{30}=\underline{\$735}$

Table 12.10 Vehicle category classification (FHWA 2001)

Vehicle category classification
1. Passenger cars and other 2-axle, 4-tired passenger vehicles
Motorcycles
Passenger cars
Other two-axle, four-tire single unit vehicles
2. Single-unit trucks, 2-axle, 4-tired or more commercial trucks
Buses
Two-axle, six-tire, single-unit trucks
Three-axle single-unit trucks
Four or more axle single-unit trucks
3. Combination-unit trucks
Four or fewer axle single-trailer trucks
Five-axle single-trailer trucks
Six or more axle single-trailer trucks
Five or fewer axle multi-trailer trucks
Six-axle multi-trailer trucks
Seven or more axle multi-trailer trucks

Deck user delay and vehicle operating costs: an average annual daily traffic (AADT) flow of approximately 4,000 vehicles is used to estimate the designed traffic flow over the proposed bridge. This is based on the AADT flow of a road in St. James, Missouri, for the year 2000 (MoDOT 2001). The percentage breakdown of traffic for passenger cars, single-unit trucks, and combination trucks (see Table 12.10 for a description of these categories) are estimated at 90%, 5.4%, and 4.6%, respectively (FHWA 1998). Given that the majority of the traffic is passenger cars, we use this category to estimate the construction costs due to user delays and vehicle operation. Also, because of this low traffic volume, it can be assumed that the bridge can be fully closed during construction operations (FHWA 2009). To estimate the cost per mile, we use the 2011 IRS Standard Mileage Rate of \$0.51 per mile (http://www.irs.gov). Using this cost and the AADT flow, the cost per day for 1.0 mile detour can be estimated:

$$\begin{aligned}\text{Cost per day for 1.0 mile detour} &= (\text{cost per mile}) \times (\text{traffic volume per hour}) \\ &= (\$0.51/\text{mile}) \times (4000\,\text{vehicles/day}) = \$2{,}040/\text{day}\end{aligned}$$

To estimate the bridge closure time necessary for construction, we use a ratio of square footage of the proposed bridge and the conventional concrete bridge in Chandler's study, which is 3.47%. The duration for the conventional deck replacement was estimated at 50 days of actual construction work, following an 8 h work day with no overtime. Because the concrete curing process time cannot be scaled, it is excluded from this average and then added to the scaled time. Multiplying the 3.47% ratio by the total number of hours of bridge closure in Chandler's study requires a 30 h closure of the proposed bridge; and adding the curing time for the bridge deck brings the total to 366 h, or 15.25 days. Assuming a half-mile detour, this brings the total bridge vehicle operating cost for the construction of the bridge deck to:

$$\begin{aligned}\text{Delay costs for 0.5 mile detour} &= (\text{cost per day per mile})\times(\text{detour})\times(\text{closure days})\\ &= (\$2{,}040/\text{day/mile})\times(0.5\text{miles})\times(15.25\text{days})\\ &= \underline{\$15{,}555}\end{aligned}$$

12.5.1.2 Deck Surfacing and Resurfacing Costs

Surfacing material: the material cost per square foot for deck surfacing was found by taking the results for the concrete materials cost (no rebar is necessary) and adjusting it for a depth of 2 in, instead of the 10 in. The 2 in concrete deck surfacing material cost per square foot is:

$$\begin{aligned}\text{Deck resurfacing cost}/\text{ft}^2 &= (\text{cost per cubic foot})\times(\text{depth of deck replaced})\\ &= (\$7.70/\text{ft}^3)\times(2/12\text{ft.}) = \$1.28/\text{ft}^2\end{aligned}$$

The total materials cost for deck resurfacing is:

$$\begin{aligned}\text{Material cost of deck resurfacing} &= (\text{deck resurfacing cost}/\text{ft}^2)\times(\text{bridge area})\\ &= (\$1.28/\text{ft}^2)\times(26.25\,\text{ft}\times 27.33\,\text{ft}) = \underline{\$921}\end{aligned}$$

Surfacing labor: no data about labor cost for deck surfacing is available, so we prorate the $9 per square foot cost assumed previously for the two-inch resurfacing concrete thickness. This is a reasonable assumption since the deck serves as formwork for the resurfacing concrete and should require less labor to construct. The total labor cost for deck resurfacing is:

$$\begin{aligned}\text{Labor cost of deck resurfacing} &= (\text{labor cost of deck replacement})\times(\text{prorate factor})\\ &= (\$6{,}457)\times(2\,\text{in}/10\,\text{in}) = \underline{\$1{,}291}\end{aligned}$$

Surfacing fuel and equipment rental: again, fuel costs are accounted for as fuel consumed and emissions released. For deck surfacing fuel consumption Table 12.11 lists general equipment needed and the cost of fuel – total fuel cost is $12,476. To account for the fuel costs, neglecting joint replacement, an approximation of 68% of the fuel consumed is used (Chandler 2004). Therefore, the total fuel cost for the deck resurfacing would be $7,236. Dividing by the square footage of the whole project amounts to $0.34 per square foot from Kendall's study, and $0.25 in Chandler's research. The average of these two values is taken – $\$0.29/\text{ft}^2$. For our study, we assume the cost of equipment rental to be 25% of the cost of fuel; therefore, we use $1.25 \times \$0.29/\text{ft}^2 = 0.37/\text{ft}^2$ for fuel and equipment rental costs.

$$\begin{aligned}\text{Fuel cost of deck resurfacing} &= (\text{fuel costs per square foot})\times(\text{bridge area})\\ &= (\$0.37\text{ft}^2)\times(26.25\text{ft}\times 27.33\text{ft}) = \underline{\$260}\end{aligned}$$

Table 12.11 Deck resurfacing/joint replacement fuel consumption (Kendall 2004)

Equipment	hp rating	Hours used	GMPH[a]	Total gallons
Crawler-mounted hydraulic excavator	428	0	12.36	0.00
Air compressor	350	48	14.56	699.00
Concrete mixer	8	0	0.23	0.00
Concrete paving machine	250	32	7.22	230.96
Concrete truck	300	32	8.66	277.15
Crane, 50 ton	177	0	5.11	0.00
Dumper	23	80	0.66	53.12
Hydraulic hammer	100	0	2.89	0.00
Motor grader	165	16	4.76	76.22
Signal boards	6	7,680	0.25	1,917.26
Vacuum truck	177	32	5.11	163.52
Water truck	450	32	12.99	415.73
Wheeled front-end loader	235	48	6.78	325.66
Fuel cost ($3 per gallon)				$12,476

[a]Gallons used per machine hour

Surfacing demolition: again, one of the end of life disposal costs is demolition of surfacing materials, including labor and fuel. These costs are taken into account for a 60-year horizon. For this study, the SRC deck would need to be resurfaced three times, which results in four bridge deck surfacing material disposals. Given that this is directly related to the material volume, we use the results of the deck installation and replacement costs determined in the last section, scaled for the smaller thickness:

$$\begin{aligned}\text{Surfacing demolition costs} &= (\text{deck demolition costs}) \times (\text{scale factor}) \\ &= (\$1,244) \times (2\,\text{in}/10\,\text{in}) = \underline{\$249}\end{aligned}$$

Surfacing disposal: only concrete is used in surfacing; and since this is directly related to the material volume, we use the results of the deck installation and replacement costs determined in the last section, excluding the revenue from the recycled steel, and scaled for the smaller two-inch thickness. The total cost for disposal adjusted for 30 years (discount rate, r of 2%) is given by the sum of the costs of landfill use, fuel, and labor, $(\$1,637+\$18+\$18)\ /\ (1+0.02)^{30}=\$5,426$, which is scaled down for the 2 in surfacing as follows:

$$\begin{aligned}\text{Surfacing disposal costs} &= (\text{deck disposal costs}) \times (\text{scale factor}) \\ &= (\$924) \times (2\,\text{in}/10\,\text{in}) = \underline{\$185}\end{aligned}$$

Surfacing user delay and vehicle operating costs: Using the cost $2,040 per day for 1.0 mile detour estimated in the deck construction section, we can estimate the delay cost for surfacing. The total time required to complete deck surfacing and joint

Table 12.12 Total user delays for detours

Vehicle type	Traffic volume	Total cost of 0.5 mile detour time
Passenger auto	142	\$45.16
Single unit truck	9	\$4.58
Combination unit truck	7	\$4.29
Total	158	\$54.03

replacement is given in Chandler as 40 days. By excluding the joint replacement activities and scaling the delay time to the proposed bridge deck size (3.47% computed in the last section), approximately 3 days are necessary to complete this activity (Table 12.12). We do not account for the concrete curing process time in this activity because early strength concrete can be used for surfacing and resurfacing. Using this delay time and assuming a half-mile detour, the total cost is estimated at:

$$\begin{aligned}\text{Delay costs for 0.5 mile detour} &= (\text{cost per day per mile})\times(\text{detour})\times(\text{closure days})\\ &= (\$2{,}040/\text{day/mile})\times(0.5\,\text{miles})\times(3\,\text{days}) = \underline{\$3{,}060}\end{aligned}$$

12.5.1.3 Maintenance Costs

The total maintenance costs for the proposed bridge deck only include the costs of materials, labor, and fuel; all other costs are very small and can be considered negligible compared to the other costs.

Maintenance material: the deck surfacing material cost per square foot found in the last section, \$1.28/ft^2 can be used for the maintenance resurfacing phase costs. To find the total square footage for repairs, a ratio between the area of repair material and the total bridge area is found using Chandler's research. The estimate of this ratio for Chandler's bridge deck can be found as follows:

$$\begin{aligned}\text{Ratio of material cost for maintenance} &= (\text{volume of repair})/(\text{depth of repair})/(\text{total bridge area in square meters})\times 100\%\\ &= (2.5\text{m}^3)/(0.0508\text{m})/(1931\text{m}^2)\times 100\% = 2.55\%\end{aligned}$$

Using this ratio we can estimate the area expected to need repairs in the proposed bridge. Multiplying 2.55% by our total bridge area (717 ft^2) yields 18.3 square feet. The cost of materials for maintenance is:

$$\begin{aligned}\text{Material cost of maintenance} &= (\text{maintenance area in sq.ft.})\times(\text{material cost per sq. ft.})\\ &= (18.28\text{ft}^2)\times(\$1.28/\text{ft}^2) = \underline{\$23}\end{aligned}$$

Maintenance labor: no data about labor cost for deck maintenance is available, so we prorate the \$9 per square foot cost assumed previously for the 18.28 ft^2 concrete maintenance area. This is a reasonable assumption since the deck serves

Table 12.13 K-A-B-C scale severity (FHWA 1994)

Severity	Description	1994	2010
K	Fatal	\$2,600,000	\$4,172,237
A	Incapacitating	\$180,000	\$288,847
B	Evident	\$36,000	\$57,769
C	Possible	\$19,000	\$30,489
Average of A–C	Injury	–	\$188,553
PDO	Property damage only	\$2,000	\$3,209

as formwork for the repair of concrete and should require less labor to construct. The total labor cost for deck maintenance is:

$$\text{Labor cost of deck maintenance} = (\text{labor cost of deck replacement}) \times (\text{prorate factor})$$
$$= (\$6{,}457) \times (18.28\,\text{ft}^2 / 717\,\text{ft}^2) = \underline{\$165}$$

Maintenance fuel and equipment rental: the fuel costs for maintenance is based on the cost of fuel consumed by the concrete mixer (see Table 12.11), which is 16 h times 0.23 GMPH times the gas price of \$3.00/ gallon, or \$11.09. Dividing by the square footage of the whole project amounts to \$0.0.00052 per square foot from Kendall's study, and \$0.00054 in Chandler's research. The average of these two values is taken – \$0.00053/ ft^2. For our study, we assume the cost of equipment rental to be 25% of the cost of fuel; therefore, we use $1.25 \times \$0.00053/\,ft^2 = \$0.00067/$ ft^2 for fuel and equipment rental costs.

$$\text{Fuel cost of maintenance} = (\text{total sq ft of bridge}) \times (\text{average cost per sq ft})$$
$$= (717\,\text{ft}^2) \times (\$0.00067 / \text{ft}^2) = \underline{\$0.48}$$

12.5.1.4 Social Costs

These are considered miscellaneous costs and include indirect costs associated with life safety and environmental effects. We concentrate on the life safety costs first, which are estimated following the analysis used by Chandler (2004).

Life safety: Costs associated with traffic accidents in construction zones are difficult to estimate because of the lack of data. In this study, we correlate the miles of construction work zone to the costs associated with property damage, personal injuries, and fatalities associated with traffic crashes. The K-A-B-C scale was used to rank each injury category as shown in Table 12.13 (FHWA 1994). This table also shows the costs associated with different types of injury category, which have been converted from 1994 to 2010 dollars using an inflation rate of 3%.

Based on 1995–1997 data from the National Highway Traffic Safety Administration (NHTSA 1996, 1997, 1998) shown in Table 12.14, an average of 92,387 people were involved in construction zone related accidents. Furthermore, Table 12.15 lists national statistics of the total number of people killed or injured,

Table 12.14 Motor vehicle injuries/fatalities in highway work zones (NHTSA 1996, 1997, 1998)

	Injuries			Fatalities			
Year	# of motor vehicle crashes	# of people injured	People injured per crash	# of fatal motor vehicle crashes	Number of deaths	Deaths per fatal crash	Property damage only
1995	26,000	42,000	1.62	665	789	1.19	56,000
1996	24,000	37,000	1.54	635	717	1.13	52,000
1997	23,000	36,000	1.57	561	658	1.17	52,000
Average	24,333	38,333	1.58	620	721	1.16	53,333

Table 12.15 Motor vehicle occupants and nonoccupants killed, injured and property damage. Only (PDO) (NHTSA 1996, 1997, 1998)

Year	Killed	Injured	PDO	Total
1995	41,817	3,465,000	4,409,000	7,915,817
1996	42,065	3,483,000	4,458,000	7,983,065
1997	43,013	3,348,000	4,542,000	7,933,013
Average	42,298	3,432,000	4,469,667	7,943,965

and "property damage only" (PDO) crashes. The average number of people involved in accidents from 1995 to 1997 was 7,943,965. By comparing the data in Tables 12.14 and 12.15, a ratio was formed relating construction zone accidents to total number of accidents for the whole country – 1.163%, which we will assume in our analysis of the proposed bridge. Assuming that the proposed bridge is located in California, we can now relate the national accident statistics to those from California. In 2004, California experienced a total of 538,295 traffic accidents (CHP 2008): 3,701 fatal traffic accidents resulting in 4,094 fatalities (0.69% of total), 203,386 injury-related collisions resulting in 302,357 injured people (37.78% of total, which is not broken-down by the A-B-C scale), and 331,208 property damage collisions (61.53% of total). So, using the 1.163% national ratio of construction zone accidents to total number of accidents for the whole country, we estimate the number of construction zone accidents in California to be 6,260 per year (0.01163 times 538,295). This result yields a total of 375,600 accidents over the 60-year horizon of the proposed bridge. Also, given that there is a total of 169,793 miles of roadway in California (FHWA 2005) and that the bridge span is 8 m (0.004971 miles), the bridge deck represents a 2.928×10^{-6}% (0.004971 divided by169,793) of all roads under construction in California. With this result, we determine an average of 0.011 construction related crashes ($2.928 \times 10^{-6}\% \times 375{,}600$) over the 60-year horizon of the proposed bridge deck. This translates to 7.59×10^{-5} (0.69% times 0.011) fatal traffic accidents, 4.15×10^{-3} (37.78% times 0.011) injury-related collisions, and 6.77×10^{-3} (61.53% times 0.011) property damage collisions. To determine the total cost, we multiply these values by the costs listed in Table 12.13 (using the average cost for A-B injury costs): \$317 for fatal (7.59×10^{-5} times \$4,172,237), \$782 for injury (4.15×10^{-3} times \$188,553), and \$22 for property damage only (6.77×10^{-3} times \$3,209). Therefore, the life cycle traffic crash cost amounts to \$1,121.

Environmental effects: To determine these costs, we use the costs associated with CO_2 emissions damage due to materials (concrete and steel) processing. The cost associated with a metric tonne of CO_2 is \$6.22 (Chandler 2004), or \$7.5/tonne adjusted for inflation; that is, \$6.22/tonne $\times (1+0.03)^6$. Also, CO_2 emissions per tonne of concrete are approximately 0.92 tonnes, and for steel approximately 0.76 tonnes per tonne of steel (see Table 12.3). Furthermore, the total volume of concrete and steel can be estimated using the results of Sect. 5.1.1 and 5.1.2. The weight of the concrete is two times (volume of concrete deck times 0.075 tons per cubic feet) plus four times (volume of concrete resurfacing times 0.075 tons per cubic feet), while the weight of the steel is two times the steel reinforcement for the deck, 6 ton: concrete weight is 125 ton, (2 × 26.25 ft × 27.33 ft × 10 in./(12 in./ft) × 0.075 ton/ft^3+4 × 26.25 ft × 27.33 ft × 2 in./(12 in./ft) × 0.075 ton/ft^3). Therefore, the total weight of CO_2 in metric tonnes is 6 ton of steel times (0.91 tonnes/ton) times 0.755 tonneCO_2/tonne plus 125 ton of concrete times (0.91 tonnes/ton) times 0.918 tonne CO_2/tonne equals 108 tonnes of CO_2; the cost of which is \$7.5/tonne CO_2 × 108 tonnes = \$814. The costs associated with transportation and construction are relatively small and are excluded from the lifecycle cost analysis.

12.5.1.5 Steel Reinforced Concrete Bridge Deck Total Cost Estimate

The total cost estimates for the SRC bridge deck alternative are summarized in Table 12.16. The costs for any future activity will be computed using present worth, which for this case can be assumed to be based on a discount rate of 2% as discussed earlier. For example, the cost of materials for Bridge Deck Installation & Replacement is obtained by adding the cost of the deck replacement at the end of 30 years to the initial cost of the deck materials; that is, total cost at present value = $C_{PV}=C_{CP}[1+(1+r)^{-n}]$, or $\$14,061\times[1+(1+0.02)^{-30}]=\$21,824$. The other Bridge Deck Installation & Replacement costs were determined following a similar approach. Also, the cost of materials for Bridge Deck Resurfacing is obtained by adding the cost of the deck resurfacing (three times at 15 years, 30 years, and 45 years) to the initial cost of the deck materials; that is, total cost at present value= $C_{PV}=C_{CP}\ [1+\Sigma(1+r)^{-n}]$, or $\$921\times[1+(1+0.02)^{-15}+(1+0.02)^{-30}+(1+0.02^{-45})]=\$2,492$. Finally, the cost of materials for Bridge Deck Maintenance is obtained by adding the cost of the deck maintenance (seven times at 5 years, 10 years, 20 years, 25 years, 35 years, 40 years, 50 years, and 55 years) to the initial cost of the deck materials; that is, total cost at present value = $C_{PV} = C_{CP}\ \Sigma(1+r)^{-n}$, or $\$23 \times [(1+0.02)^{-5} + (1+0.02)^{-10} + (1+0.02)^{-20} + (1+0.02)^{-25} + (1+0.02)^{-35} + (1+0.02)^{-40} + (1+0.02)^{-50} + (1+0.02)^{-55}] = \$130$. No initial costs are associated with maintenance.

We also conducted a sensitivity analysis looking at the impact of the discount rate on the total cost; the range of values for the discount rate are from −2% to 6%. A discount rate of 4% was used in Chandler's (2004) study and other discount rates have been reported in the literature (Chen and Lui 2005). Figure 12.7 shows the results. The curve's asymptote is \$48,000; that is, when the discount rate approaches infinity, all future costs approach zero.

Table 12.16 Steel reinforced concrete bridge deck cost estimate

Description	Unit cost	Frequency	Total cost
Bridge deck installation & replacement costs			
Materials	$14,061	2	$21,824
Labor	$6,457	2	$10,022
Fuel and equipment rental	$1,794	2	$2,784
Total demolition	$1,244	2	$1,931
Net disposal	$735	2	$1,141
Vehicle operating costs (0.5 miles detour) including delay	$15,555	2	$24,142
Bridge deck resurfacing costs			
Materials	$921	4	$2,492
Labor	$1,291	4	$3,493
Fuel and equipment rental	$260	4	$703
Total demolition	$249	4	$674
Total disposal	$185	4	$500
Vehicle operating costs (0.5 miles detour) including delay	$3,060	2	$8,278
Bridge deck maintenance costs			
Materials	$23	8	$130
Labor	$165	8	$935
Fuel and equipment rental	$0.48	8	$3
Miscellaneous costs			
Traffic accidents	$1,121	1	$1,121
Environmental	$814	1	$814
Total costs			$80,987

12.5.2 Fiber Reinforced Polymer Bridge Deck Research Methodology

Table 12.17 lists a number of FRP deck constructions. The Duraspan® bridge deck fabricated by Martin Marietta Composites has been used 27 times – the most of any system. Therefore, we will use this bridge deck model for this study. In order to compare this alternative to the SRC deck bridge, a lifecycle cost analysis is also conducted. Again, this lifecycle cost analysis is based on costs from materials, transportation, construction (labor, equipment, fuel, and miscellaneous costs), maintenance, end of life disposal (labor, equipment, fuel, and miscellaneous costs), environmental (primarily CO_2), and social costs (traffic congestion, delays, and accidents). As noted in the last section, FRP bridges generally have higher initial costs compared to conventional concrete; however, FRP decks require less maintenance, and can be constructed faster, which may lead to a lifecycle cost smaller than for SRC bridges. For this part of the study, we only focus deck construction and maintenance. Deck construction is even more expensive than the SRC bridge deck. Again, the lifecycle cost analysis presented here will cover each stage of the life of a bridge deck, from manufacturing to end of life disposal (Fig. 12.6).

Table 12.17 FRP bridge decks and descriptions (Telang et al. 2006)

Manufacturer	Deck description	Number of bridges	Deck cross section
Kansas Structural Composites, Inc. (KSCI)	Sandwich-type deck with top load bearing skin, bottom sheet skin, and a deep corrugated core	12 installations; first in 1996, most recent in 2003	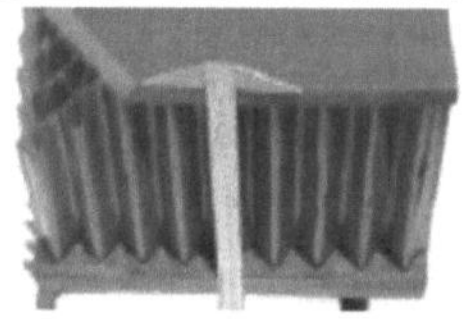
Infrastructure Composites, Inc. (ICI)	Same as the Kansas Structural Composites deck	One installation in the U. S. and one in Europe	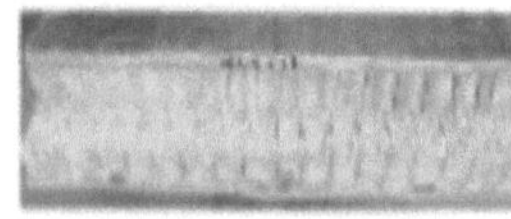
Martin Marietta Composites, Inc. (MMC)	The Duraspan deck system consists of a trapezoid cross-sectional piece manufactured by pultrusion. A Duraspan deck is post-assembled to delivered width by bonding unit pieces with epoxy or urethane adhesive.	27 installations; first in 1996, most recent completed in 2004	
Hardcore Composites, Inc. (HCI)	HCI uses various forms of Vacuum-Assisted Resin Transfer Molding (VARTM) technology to producing FRP decks. The majority of decks use a vertical standing foam box as core, the pattern for which can vary substantially.	26 installations; first in 1997, most recent in 2002	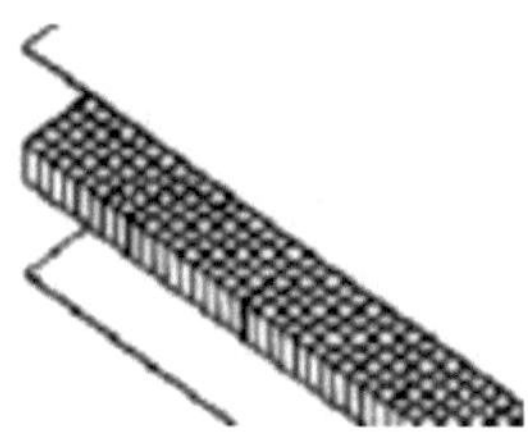
Creative Pultrusions, Inc.	The Superdeck is made with pultruded hexagonal sections bonded to form the desired deck width and is similar to the DuraSpan deck	Nine installations; first in 1997, most recent in 2002	
Strongwell, Inc.	Strongwell does not directly market vehicular decks, but many pilot composite bridge and deck projects have used Strongwell's pultruded components	Estimated three installations; earliest in 1995 and latest in 2003	

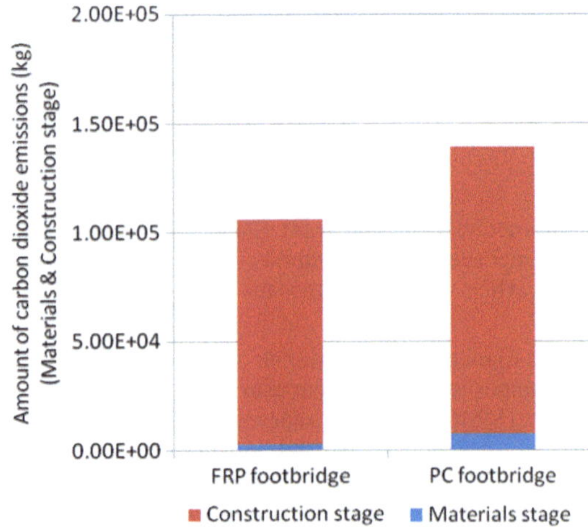

Fig. 12.6 Total carbon dioxide emissions at the materials and construction stage of FRP and PC footbridges

12.5.2.1 Installation Costs

These costs can be obtained from available data of previous construction projects, or by a detailed analysis of the costs associated with each stage of the initial construction processes. In this section, we include estimates of costs associated with materials and construction (labor and fuel).

Deck materials: The initial cost of materials for FRP bridge decks has been estimated at \$70 per square foot by Nystrom et al. (2003) and \$170 per square foot by Daniel (2003). This range of values will be used to estimate the cost of the proposed bridge deck. We also assume that any cost increase due to inflation can be offset by development of the technology as more decks are being constructed using FRP. Therefore, given that the total area of the proposed bridge was calculated earlier as 717 ft^2 (66.6 m^2), the range of cost for the proposed system is:

$$\begin{aligned}\text{Material cost of deck replacement} &= (\text{materials costs range per square foot}) \times (\text{bridge area}) \\ &= (\$70/\text{ft}^2 \text{ to } 170/\text{ft}^2) \times (717\,\text{ft}^2) = \$50{,}190 \text{ to } \$121{,}890\end{aligned}$$

Deck labor: the cost of labor used in Chandler's (2004) study was \$9 per square foot and will again be assumed for our purposes as well, which is reasonable given that the main difference in construction between the two systems will come from the decrease in construction time, so the same labor cost per square foot for concrete and FRP will be adjusted for the shorter construction time. From several successful FRP bridge deck installations (FHWA 2009), these types of projects can be completed in the range of one-third to one-half the time than of SRC bridge decks. Given that there will continue to be advances in the installation of FRP bridge decks, we can assume the lower range, one-third the time that of SRC bridge decks. Therefore, the labor cost is divided by three, or \$3 per square foot:

$$\begin{aligned}\text{Labor cost of deck replacement} &= (\text{labor costs per square foot}) \times (\text{bridge area}) \\ &= (\$3/\text{ft}^2) \times (717\,\text{ft}^2) = \underline{\$2{,}152}\end{aligned}$$

Table 12.18 FRP bridge deck installation fuel consumption

	hp rating	Hours used	GMPH[a]	Total gallons
Crane, 50 ton	177	176	5.11	899
Signal boards	6	18,000	0.25	4,494
Fuel cost ($3 per gallon)			16.08	$16,179
Cost per hour				$0.76

[a]Gallons used per machine hour

Deck fuel and equipment rental: construction machinery fuel consumption is again accounted for in terms of fuel consumed and emissions released. The equipment considered for the cost of fuel consumption was a 50 ton crane used to put the panels in place and signal boards to notify traffic of the construction work zone. The 50 ton crane was used because more specific data was available, including the hp rating and hours used for a SRC bridge deck project. The cost estimated Table 12.18 lists general equipment needed for construction of a FRP bridge deck using rates from Kendall's study for concrete in 2004. Again, we assume a fuel cost of $3 per gallon for both gas and diesel for the West Coast (EIA 2010). A total fuel cost of $0.76 per hour is used for the deck replacement phase of the bridge construction. For the FRP bridge deck the construction time was estimated at a third of that of the SRC bridge deck (366 h), as discussed earlier (FHWA 2009). This yields a total of 122 h. We again assume the cost of equipment rental to be 25% of the cost of fuel; therefore, we use 1.25 × $0.76/h = $0.95/h for fuel and equipment rental costs.

$$\text{Fuel cost of deck installation} = (\text{fuel costs per hour}) \times (\text{construction time})$$
$$= (\$0.95\,/\,\text{hour}) \times (122\,\text{hours}) = \underline{\$116}$$

Deck demolition: one of the end of life disposal costs is demolition, including labor and fuel.

Demolition costs were taken into account throughout the lifecycle of the FRP bridge, which in this case is at the end of the 60-year bridge life. Given that FRP bridge decks are lightweight and approximately 20% of the weight of an equivalent SRC bridge deck (Nelson 2005), we assume the equipment use cost associated with demolition is only 20% that of the SRC bridge deck, or 0.2($1,126) = $225. However, demolition will not occur until 60 years from now; therefore, we must discount this amount to current value by using the discount rate factor, $\$225/(1+r)^n = \$225/(1+0.02)^{60} = \$69$, assuming $n = 60$ years and a discount rate, r of 2%. Also, in this case, labor is directly linked to equipment use; therefore, labor and equipment costs should be proportional to fuel costs, and we assume that equipment rental and labor costs are equal to the cost of fuel, $69. The total demolition cost is <u>$138</u>.

Deck disposal: we assumed that all of this material is diverted to a landfill due to lack of information about FRP recycling. Earlier we established the median landfill tipping fee at $36.5/ton for California (CalRecycle 2010). We use this value for disposal of materials. Assuming FRP mass equals one-fifth that of the SRC bridge deck, the mass of FRP removed from the bridge deck can be calculated by taking one-fifth the weight of the concrete calculated earlier, or 44.8 ton/5 = 9 ton. The cost of landfill disposal is:

$$\text{Landfill tipping costs} = (\text{mass of waste FRP}) \times (\text{fee/ton})$$
$$= (9\ \text{tons}) \times (\$36.50/\text{ton}) = \$328$$

Given that the weight capacity of a truck to transport the FRP is 24 ton (Chandler 2004), only one truck is needed to transport the demolished FRP to the landfill. According to Bradley (2000), dump trucks are classified as class 8 vehicles. Vehicles in this class have a fuel efficiency of 6.6 mpg. Assuming a distance of 20 miles to the disposal site is assumed and a cost of fuel of \$3/gallon, we get the transportation cost as follows:

$$\begin{aligned}\text{Transportation fuel costs} &= (\text{truck loads}) \times (\text{distance to landfill}) / (\text{fuel efficiency}) \\ \text{for concrete demolition} &\quad \times (\text{cost of fuel}) \\ &= (1\ \text{load}) \times (20\ \text{miles}) / (6.6\ \text{mpg}) \times (\$3.00/\text{gal.}) = \$9\end{aligned}$$

In this case, labor is also directly linked to equipment use, labor and equipment costs should be proportional to fuel costs; we will assume that equipment rental and labor costs are equal to the cost of fuel. That is, \$9.

The total cost for disposal adjusted for a payout in 60 years is given by the sum of the costs of landfill use, fuel, and labor adjusted using $n = 60$ years and a discount rate, i of 2%.; that is, $(\$328 + \$9 + \$9)/(1 + 0.02)^{60} = \underline{\$106}$.

Deck user delay and vehicle operating costs: for the FRP bridge deck, the construction time was estimated at a third of that of the SRC bridge deck (366 h). This yields a total of 122 h or 5.1 days. Using the same parameters as for the SRC bridge deck, the construction costs due to user delays and vehicle operation for the FRP bridge deck should be a third. The total bridge vehicle operating cost for the deck installation 0.5 mile detour is \$5,185.

12.5.2.2 Maintenance Costs

The innovation of FRP material and its application to bridge decks has led to limited maintenance issues, which makes it impossible to collect performance data and maintenance records (Hong et al. 2006 and Hong and Hastak 2006). Therefore, maintenance costs for FRP bridge decks are taken as \$6.51 per m^2 based on NCDOT standards (Nystrom et al. 2003).

$$\text{Cost of maintenance} = (\text{maintenance area in sq.meters}) \times (\text{cost per sq.meters})$$
$$= (8.3\ \text{m} \times 8.0\ \text{m}) \times (\$6.51/\text{m}^2) = \underline{\$432}$$

12.5.2.3 Miscellaneous Costs

Life safety: the 0.011 crashes over the 60-year time frame of the SRC bridge deck will be cut by two-thirds because of the faster installation time of the FRP bridge deck. The faster installation time decreases the probability of accidents occurring on the bridge because construction time is reduced. This amounts to a cost of $374 associated with traffic accidents.

Environmental effects: to determine these costs, we use the costs associated with CO_2 emissions damage due to FRP materials processing. The cost associated with a metric tonne of CO_2 is $6.22 (Chandler 2004), or $7.5/tonne adjusted for inflation, as determined earlier. Also, CO_2 emissions per tonne of FRP are approximately 3.09 tonnes since our system is fabricated by pultrusion (see Table 12.3). Here we use the assumption that the weight of FRP materials is one-fifth that of the concrete materials. The weight of the concrete is volume of concrete deck times 0.075 ton per cubic feet, 0.2 × 26.25 ft × 27.33 ft × 10 in/(12 in/ft) × 0.075 ton/ft^3, or 9 ton of FRP materials. Therefore, the total weight of CO_2 in metric tonnes is 9 tons of FRP times (0.91 tonnes/ton) times 3.09 tonneCO$_2$/tonne equals 28 tonnes of CO_2; the cost of which is $7.5/tonne CO_2 × 28 tonnes = $208.

12.5.2.4 Fiber Reinforced Polymer Bridge Deck Total Cost Estimate

Table 12.19 lists a cost estimate for the FRP bridge deck alternative. The total cost estimate using the FRP material ranges from $58,901 to $130,601, or $82 to $182 per square foot. So, assuming a 2% discount rate, the breakeven cost for FRP materials is $113 per square foot for the FRP alternative to equal the SRC alternative at a discount rate of 2%. This is below the FRP range average cost of $132 per square foot. It is interesting to note that the lower extremity of the FRP cost range, $82 per square foot is lower than the $93 per square foot cost for a SRC bridge deck assuming a 4% discount rate, see Fig. 12.7.

12.6 Conclusions and Recommendations

In this chapter, a detailed study of the economic and environmental costs of using FRP composites to construct a short-span bridge was performed. Where data was unavailable, assumptions and generalizations were made that may lead to conclusions, which might not be applicable to systems other than those similar to the ones presented in this study.

Table 12.19 Fiber reinforced polymer bridge deck cost estimate

Description	Unit cost	Frequency	Total cost
FRP bridge deck installation costs			
Materials	$50, 190 to $121,890	1	$50, 190 to $121,890
Labor	$2,152	1	$2,152
Fuel and equipment rental	$116	1	$116
Total demolition	$138	1	$138
Net disposal	$106	1	$106
Vehicle operating costs (0.5 miles detour) including delay	$5,185	1	$5,185
FRP bridge deck maintenance costs			
Maintenance	$432	1	$432
Miscellaneous costs			
Traffic accidents	$374	1	$374
Environmental	$208	1	$208
Total costs			$58,901–$130,601

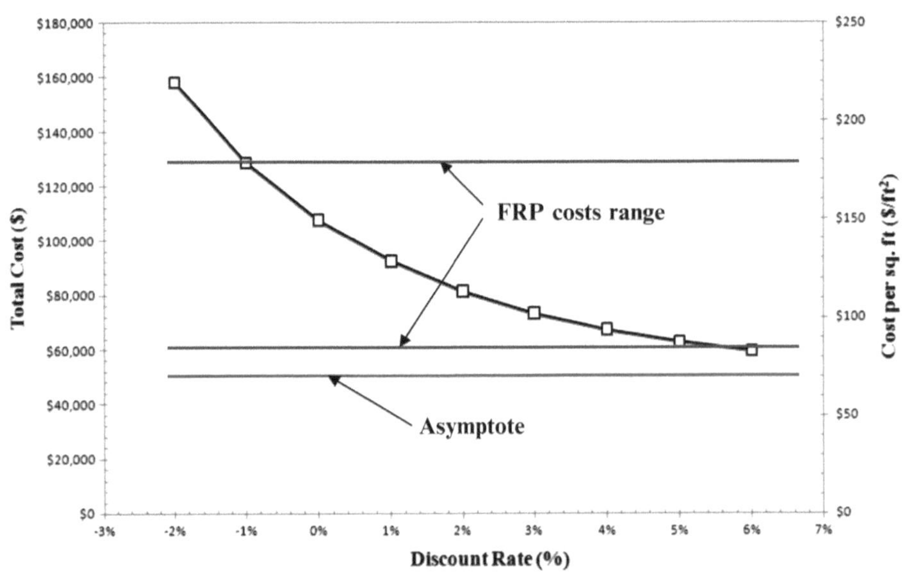

Fig. 12.7 Results of the effect of discount rate on the total cost

12.6.1 Conclusions

FRP outperformed SRC in all stages of the lifecycle costs including the overall lifecycle cost with the lower extremity for FRP material cost for a discount rate of 2%. The upper extremity cost for FRP material is much higher because of the initial installation material cost, which results in a higher total lifecycle cost. The breakeven point found in this study for the cost of FRP material compared to the SRC bridge deck at 2% discount rate is $113 per square foot. This value varies depending on the discount rate as shown by the sensitivity analysis depicted in Fig. 12.7.

12.6.2 Recommendations

One recommendation would be to incorporate actual costs from previous studies to gain closer approximation of the FRP bridge deck costs. The large range of potential costs introduces inaccuracy within the data resulting in skewed results. Although guidelines have been developed, there exists a need for code-based procedures and design criteria for life cycle assessment before widespread adoption of FRP materials is fully realized.

12.6.3 Future Studies

FRP materials are still in intermediate stages of development; as the use of FRP materials becomes more widespread in civil infrastructure, it's likely that material costs will decrease. Also, future studies should broaden the scope for life cycle costs to include more detailed environmental cost considerations; including air and water pollution. Different materials should be tested, including natural fiber composites that have lower environmental impacts. Some candidate materials should include hemp, china reed, and flax fibers, which have been shown to yield a further reduction in weight when compared to glass fibers (and are recyclable), resulting in recovered energy and carbon credits (Joshi et al. 2003). This is particularly applicable for use with short span bridges because of lower AADT.

The scope of this study pertained to short-span bridges, but should be expanded to encompass medium and long span bridges. These bridges are characterized by higher traffic volumes and longer construction periods. The longer duration required for construction of long-span bridges will maximize the benefit of FRP decks, primarily due to shorter construction period and minimal use of heavy construction equipment, resulting in greater savings in a lifecycle cost analysis.

References

Anderson J, Jansz A, Steele K, Thistlethwaite P, Bishop G, Black A (2004) Green guide to composites: an environmental profiling system for composite materials and products. Building Research Establishment (BRE) and NetComposites, Chesterfield, USA

ASCE (2009) Report card for America's infrastructure. American society of civil engineers. Bridges. http://www.infrastructurereportcard.org/fact-sheet/bridges. Accessed 2 June 2010

Biondini F, Frangopol DM (eds) (2008) Life-cycle civil engineering. CRC Press/Balkema, Taylor & Francis group, Londan, UK
Bradley R (2000) Technology roadmap for the 21st century truck program. U.S. Department of Energy: Energy Efficiency and Renewable Energy, report no 21CT–001 Oak Ridge, TN
CalRecycle (2010) Recycling rate for mid valley. http://www.calrecycle.ca.gov/Profiles/Facility/Landfill/. Accessed 8 Oct 2010
Caltrans, California Department of Transportation (2009) Construction statistics 2009. http://www.dot.ca.gov/hq/esc/estimates/Construction_Stats_2009.pdf. Accessed 14 Feb 2011
Caltrans, California Department of Transportation (2010) Average highway contract prices. http://www.dot.ca.gov/hq/esc/oe/contract_progress/exhibitBE.pdf. Accessed 14 Feb 2011
Chandler RF (2004) Life-cycle cost model for evaluating the sustainability of bridge decks, Center for sustainable systems, University of Michigan, CSS04-06, 1–116
Chen W, Lui EM (2005) Handbook of structural engineering, 2nd edn. CRC Press, Boca Raton
CHP, California Highway Patrol (2008) 2008 Annual report of fatal and injury motor vehicle traffic collisions. http://www.chp.ca.gov/switrs/. Accessed 10 June 2010
Daniel R A (2003) Environmental considerations to structural material selection for a bridge. In: Proceedings of the European bridge engineering conference: lightweight bridge decks, Rotterdam, Mar 2003
Ehlen MA (1999) Life-cycle costs of fiber-reinforced-polymer bridge decks. J Mater Civ Eng 11:224–230
EIA (2010) U.S. energy information administration gas and diesel rates. http://www.eia.doe.gov/oog/info/gdu/gasdiesel.asp
FHWA, Federal Highway Administration (1994) Motor vehicle accident costs. http://safety.fhwa.dot.gov/facts_stats/t75702.cfm. Accessed 14 July 2010
FHWA, Federal Highway Administration (1998) Life-cycle cost analysis in pavement design. http://isddc.dot.gov/OLPFiles/FHWA/013017.pdf. Accessed 30 June 2010
FHWA, Federal Highway Administration (2001) Vehicle classification ,monitoring. http://www.fhwa.dot.gov/ohim/tmguide/tmg4.htm#app4c. Accessed 14 July 2010
FHWA, Federal Highway Administration (2005) Public road length – 2004. http://www.fhwa.dot.gov/policy/ohim/hs04/htm/hm10.htm. Accessed 30 June 2010
FHWA, Federal Highway Administration (2009) Prefabricated bridge elements and systems cost study: accelerated bridge construction success stories. http://www.fhwa.dot.gov/bridge/prefab/successstories/091104/01.cfm. Accessed 8 July 2010
Hong T, Hastak M (2006) Construction, inspection, and maintenance of FRP deck panels. J Compos Constr 10:561–572
Hong T, Han S, Lee S (2006) Simulation-based determination of optimal life-cycle cost for FRP bridge deck panels. Autom Constr 16:140–152
Joshi SV, Drzal LT, Mohanty AK, Arora S (2003) Are natural fiber composites environmentally superior to glass fiber reinforced composites? Compos Part A: Appl Sci Manuf 15:371–376
Kendall A (2004) A Dynamic life cycle assessment tool for comparing bridge deck designs. Center for sustainable systems, University of Michigan, report no CSS04-12. http://css.snre.umich.edu/css_doc/CSS04-12.pdf. Accessed 2 Feb 2011
Mahadvi A, Ries R (1998) Towards computational eco-analysis of building designs. Comput Struct 67:375–387
Martin Marietta Composites (2004) Broadway bridge installation marks historic milestone. The conveyer. http://www.martinmarietta.com/Products/BroadwayBridge.pdf. Accessed 14 July 2010
MoDOT, Missouri Department of Transportation (2001) District 9 2000 traffic volume counts. http://www.modot.mo.gov/safety/documents/D9TrafVol.pdf. Accessed 14 July 2010
Mu B, Wu H, Yan A, Warnemuende K, Fu G, Gibson RF, Kim D (2006) FEA of complex bridge system with FRP composite deck. J Compos Constr 10:79–86
Nelson J L (2005) Behavior of GFRP bridge decks for highway bridges. MS thesis, civil engineering, North Carolina State University, Sept 2005

NHTSA, National Highway Traffic Safety Administration (1996) Traffic safety facts 1995. http://www.nhtsa.gov/people/ncsa/FactPrev/overvu95.html. Accessed 30 June 2010
NHTSA, National Highway Traffic Safety Administration (1997) Traffic safety facts 1996. http://www.nhtsa.gov/people/ncsa/TSF96.html. Accessed 30 June 2010
NHTSA, National Highway Traffic Safety Administration (1998) Traffic safety facts 1997. http://www.nhtsa.gov/people/ncsa/pdf/TSF97.pdf. Accessed 30 June 2010
Nystrom HE, Watkins SE, Nanni A, Murray S (2003) Financial viability of fiber-reinforced polymer (FRP) bridges. J Manag Eng 19:2–8
Roxsand (2007) Products and services. http://roxsand.net/products.php. Accessed 8 Oct 2010
Shimomura T, Nishizaki I, Tanaka H (2009) Report on JSCE committee activities on LCA and LCC of FRP infrastructure. In: Proceedings of US-Japan workshop on life cycle assessment of sustainable infrastructure materials, Sapporo
Telang NM, Dumlao C, Mehrabi AB, Ciolko A T, Gutierrez J (2006) Field inspection of in-service FRP bridge decks. National Cooperative Highway Research Program (NCHRP). Transportation Research Board http://onlinepubs.trb.org/onlinepubs/nchrp/nchrp_rpt_564.pdf. Accessed 12 June 2010
van Oss HG (2010) Cement. United States Geological Survey, 2008 minerals yearbook, vol 16. pp. 1–38

Abbreviations

AADT	Annual Average Daily Traffic	12
AAR	Alkali-Aggregate Reactions	11
AASHTO	American Association of State Highway and Transportation Officials	9
ACEC	American Council of Engineering Companies	2
ACI	American Concrete Institute	1
ACSIM	Army Chief of Staff for Installation Management	10
ADT	Average Daily Traffic	9
AF&PA	American Forest & Paper Association	2
AISC	American Insitute of Steel Construction	2
AISI	American Iron and Steel Insitiute	2
ANSI	American National Standard	2
APWA	American Public Works Association	2
ASCE	American Society of Civin Engineers	2
ASD	Allowable Stress Design	2
ASHRAE	American Society of Heating, Refrigerating, and Air-Conditioning Engineers	2
ASI	Australia Steelwork International	2
ASR	Alkali-Silica Reactions	11
BDI	Bridge Diagnostics Incorporated	11
BIM	Building Information Modeling	2
BOF	Basic Oxygen Furnace	2
CC	Conventional Concrete	12
CFRP	Carbon Fiber Reinforced Polymers	3
CIPP	Curled-in-place-pipe	1
CISC	Canada International Steelwork Contractors	2
CLPT	Classical Laminated Plate Theory	4
CPC	Corrosion Prevention and Control Program	1
CSI	Campbell Scientific	11
DEAM	Database for Environmental Analysis and Management	12

R. Jain and L. Lee (eds.), *Fiber Reinforced Polymer (FRP) Composites for Infrastructure Applications*, Strategies for Sustainability, DOI 10.1007/978-94-007-2357-3,

DOD	The Department of Defense	1
EAF	Electric-Air Furnace	2
ECC	Engineered Cementitious Composites	12
EW	Engineering Wood	2
FEA	Finite Element Analysis	3
FEM	Finite Element Modeling	9
FFT	Fast Fourier Transform	7
FG	Fiberglass	10
FHWA	Federal Highway Administration	2
FOS	Fiber Optic Sensors	11
FRP	Fiber Reinforced Polymer Composites	1
FSC	Forest Stewardship Council	2
GBI	Green Building Initiative	2
GFRP	Glass Fiber Reinforced Polymer	5
HDPE	High Density Polyethylene	1
HSS	Hollow Structural Shapes	2
IBC	International Building Code	2
ICC	International Codes Council	2
IGCC	International Green Construction Code	2
ILSS	Interlaminar Shear Strength	3
IMPB	Immiscible Polymer Blend Composites	10
ISCG	International Steelwork Contractors Group	2
ITTP	Installation Technology Transition Program	10
LCA	Life Cycle Assessment	2
LCC	Life Cycle Cost	12
LEED	Leadership in Energy and Environmental Design	2
LMI	Logistics Management Institute	10
LRFD	Load and Resistance Factor Deign	2
MoDOT	Missouri Department of Transportation	9
NCDOT	North Carolina State Department of Transportation	10
NEFMAC	New Fiber Composite Material for Reinforcing Concrete	8
NGBS	National Green Building Standard	2
ODOT	Ohio Department Transportation	11
OSD	Office of the Secretary of Defense	1
PBD	Preformance-Based Design	2
PC	Pre-Stressed Concrete	12
PDO	Property Damage Only	12
PP	Polypropylene	10
PS	Polystyrene	10
PTFE	Polytetrafl Uoroethylene	1
RH	Relative Humidity	3
RIM	Resin Infusion Molding	4
ROI	Return on Investment	10
RPL	Recycled Plastic Lumber	10
RT	Room Temperature	5

RTCL	Reinforced Thermoplastic Composite Lumber	10
RTM	Resin Transfer Molding	12
RVE	Representative Volume Element	3
SAISC	South Africa International Steelwork Contractors	2
SBST	Short Beam Shear Test	3
SCNZ	Steelwork Contractors of New Zealand	2
SEET	Strain Energy Equivalence Theory	10
SEI	Structural Engineering Institute	2
SEM	Scanning Electron Microscopy	10
SHM	Structural Health Monitoring System	11
SHPB	Split Hopkinson Pressure Bar	5
SIP	Stay-in-place	8
SRC	Steel Reinforced Concrete	12
SRP	Steel-Reinforced Polymer	9
SSTG	Steel Sustainability Task Group	2
UC	UV Radiation + Condensation	5
USACERL	U.S. Army Construction Engineering Research Laboratory	10
USGBC	U.S. Greeb Building Council	2
UV	Ultraviolet	1
VARIM	Vacuum Assisted Resin Infusion Molding	5
VARTM	Vacuum-Assisted Resin Transfer Molding	3
VW	Vibrating Wire	11

Index

R. Jain and L. Lee (eds.), *Fiber Reinforced Polymer (FRP) Composites for Infrastructure Applications*, Strategies for Sustainability, DOI 10.1007/978-94-007-2357-3,

P

R

S

U

V

MIX
Papier aus verantwortungsvollen Quellen
Paper from responsible sources
FSC® C105338

If you have any concerns about our products,
you can contact us on
ProductSafety@springernature.com

In case Publisher is established outside the EU,
the EU authorized representative is:
Springer Nature Customer Service Center GmbH
Europaplatz 3, 69115 Heidelberg, Germany

Printed by Libri Plureos GmbH
in Hamburg, Germany